A Multigrid Tutorial

A Multigrid Tutorial

Second Edition

William L. Briggs
Department of Mathematics
University of Colorado at Denver
Denver, Colorado

Van Emden Henson
Center for Applied Scientific Computing
Lawrence Livermore National Laboratory
Livermore, California

Steve F. McCormick
Department of Applied Mathematics
University of Colorado at Boulder
Boulder, Colorado

Society for Industrial and Applied Mathematics

Portions of this work were performed under the auspices of the U.S. Department of Energy by University of California Lawrence Livermore National Laboratory under contract W-7405-Eng-48.

Library of Congress Cataloging-in-Publication Data

Briggs, William L.
    A multigrid tutorial.—2nd. ed. / William L. Briggs, Van Emden Henson, Steve F. McCormick.
        p. cm.
    Includes bibliographical references and index.
    ISBN 0-89871-462-1 (pbk.)
    1. Differential equations, Partial—Numerical solutions. 2. Multigrid methods (Numerical analysis) I. Henson, Van Emden. II. McCormick, S. F. III. Title.

QA377.B75 2000
515'.353--dc21   00-024103

To the ladies…

Julie, Teri, Lynda, Katie, Jennifer

Contents

Preface to the Second Edition

Twelve years have passed since the publication of the first edition of *A Multigrid Tutorial*. During those years, the field of multigrid and multilevel methods has expanded at a tremendous rate, reflecting progress in the development and analysis of algorithms and in the evolution of computing environments. Because of these changes, the first edition of the book has become increasingly outdated and the need for a new edition has become quite apparent.

With the overwhelming growth in the subject, an area in which I have never done serious research, I felt remarkably unqualified to attempt a new edition. Realizing that I needed some help, I recruited two experts to assist with the project. Steve McCormick (Department of Applied Mathematics, University of Colorado at Boulder) is one of the original researchers in the field of multigrid methods and the real instigator of the first edition. There could be no better collaborator on the subject. Van Emden Henson (Center for Applied Scientific Computing, Lawrence Livermore National Laboratory) has specialized in applications of multigrid methods, with a particular emphasis on algebraic multigrid methods. Our collaboration on a previous SIAM monograph made him an obvious choice as a co-author.

With the team in place, we began deliberating on the content of the new edition. It was agreed that the first edition should remain largely intact with little more than some necessary updating. Our aim was to add a roughly equal amount of new material that reflects important core developments in the field. A topic that probably should have been in the first edition comprises Chapter 6: FAS (Full Approximation Scheme), which is used for nonlinear problems. Chapter 7 is a collection of methods for four special situations that arise frequently in solving boundary value problems: Neumann boundary conditions, anisotropic problems, variable-mesh problems, and variable-coefficient problems. One of the chief motivations for writing a second edition was the recent surge of interest in algebraic multigrid methods, which is the subject of Chapter 8. In Chapter 9, we attempt to explain the complex subject of adaptive grid methods, as it appears in the FAC (Fast Adaptive Composite) Grid Method. Finally, in Chapter 10, we depart from the predominantly finite difference approach of the book and show how finite element formulations arise. This chapter provides a natural closing because it ties a knot in the thread of variational principles that runs through much of the book.

There is no question that the new material in the second half of this edition is more advanced than that presented in the first edition. However, we have tried to create a safe passage between the two halves, to present many motivating examples, and to maintain a tutorial spirit in much of the discourse. While the first half of the book remains highly sequential, the order of topics in the second half is largely arbitrary.

The FAC examples in Chapter 9 were developed by Bobby Philip and Dan Quinlan, of the Center for Applied Scientific Computing at Lawrence Livermore National Laboratory, using AMR++ within the Overture framework. Overture is a parallel object-oriented framework for the solution of PDEs in complex and moving geometries. More information on Overture can be found at http://www.llnl.gov/casc/Overture.

We thank Irad Yavneh for a thorough reading of the book, for his technical insight, and for his suggestion that we enlarge Chapter 4. We are also grateful to John Ruge who gave Chapter 8 a careful reading in light of his considerable knowledge of AMG. Their suggestions led to many improvements in the book.

Deborah Poulson, Lisa Briggeman, Donna Witzleben, Mary Rose Muccie, Kelly Thomas, Lois Sellers, and Vickie Kearn of the editorial staff at SIAM deserve thanks for coaxing us to write a second edition and for supporting the project from beginning to end. Finally, I am grateful for the willingness of my co-authors to collaborate on this book. They should be credited with improvements in the book and held responsible for none of its shortcomings.

Bill Briggs
November 15, 1999
Boulder, Colorado

Preface to the First Edition

Assuming no acquaintance with the subject, this monograph presents the essential ideas that underlie multigrid methods and make them work. It has its origins in a tutorial given at the Third Copper Mountain Conference on Multigrid Methods in April, 1987. The goal of that tutorial was to give participants enough familiarity with multigrid methods so that they could understand the following talks of the conference. This monograph has been written in the same spirit and with a similar purpose, although it does allow for a more realistic, self-paced approach.

It should be clear from the outset that this book is meant to provide a basic grounding in the subject. The discussion is informal, with an emphasis on motivation before rigor. The path of the text remains in the lowlands where all of the central ideas and arguments lie. Crossroads leading to higher ground and more exotic topics are clearly marked, but those paths must be followed in the Suggested Reading and the Exercises that follow each chapter. We hope that this approach will give a good perspective of the entire multigrid landscape.

Although we will frequently refer to *the* multigrid method, it has become clear that multigrid is not a single method or even a family of methods. Rather, it is an entire approach to computational problem solving, a collection of ideas and attitudes, referred to by its chief developer Achi Brandt as *multilevel methods*.

Originally, multigrid methods were developed to solve boundary value problems posed on spatial domains. Such problems are made discrete by choosing a set of grid points in the domain of the problem. The resulting discrete problem is a system of algebraic equations associated with the chosen grid points. In this way, a physical grid arises very naturally in the formulation of these boundary value problems.

More recently, these same ideas have been applied to a broad spectrum of problems, many of which have no association with any kind of physical grid. The original multigrid approach has now been abstracted to problems in which the grids have been replaced by more general levels of organization. This wider interpretation of the original multigrid ideas has led to powerful new techniques with a remarkable range of applicability.

Chapter 1 of the monograph presents the model problems to which multigrid methods were first applied. Chapter 2 reviews the classical iterative (relaxation) methods, a firm understanding of which is essential to the development of multigrid concepts. With an appreciation of how the conventional methods work and why they fail, multigrid methods can be introduced as a natural remedy for restoring and improving the performance of the basic relaxation schemes. Chapters 3 and 4 develop the fundamental multigrid cycling schemes and discuss issues of implementation, complexity, and performance. Only in Chapter 5 do we turn to some theoretical questions. By looking at multigrid from a spectral (Fourier mode) point of view and from an algebraic (subspace) point of view, it is possible to give an explanation of why the basic multigrid cycling scheme works so effectively.

Not surprisingly, the body of multigrid literature is vast and continues to grow at an astonishing rate. The Suggested Reading list at the end of this tutorial [see the bibliography in the Second Edition] contains some of the more useful introductions, surveys, and classical papers currently available. This list is hardly exhaustive. A complete and cumulative review of the technical literature may be found in the *Multigrid Bibliography* (see Suggested Reading), which is periodically updated. It seems unnecessary to include citations in the text of the monograph. The ideas presented are elementary enough to be found in some form in many of the listed references.

Finally, it should be said that this monograph has been written by one who has only recently worked through the basic ideas of multigrid. A beginner cannot have mastered the subtleties of a subject, but often has a better appreciation of its difficulties. However, technical advice was frequently necessary. For this, I greatly appreciate the guidance and numerous suggestions of Steve McCormick, who *has* mastered the subtleties of multigrid. I am grateful to John Bolstad for making several valuable suggestions and an index for the second printing. For the fourth printing the Suggested Reading section has been enlarged to include six recently published books devoted to multigrid and multilevel methods. A genuinely new development is the creation of mg-net, a bulletin board/newsgroup service which is accessible by sending electronic mail to `mgnet@cs.yale.edu`. For the real production of this monograph, I am grateful for the typing skills of Anne Van Leeuwen and for the editorial assistance of Tricia Manning and Anne-Adele Wight at SIAM.

Chapter 1

Model Problems

Multigrid methods were originally applied to simple boundary value problems that arise in many physical applications. For simplicity and for historical reasons, these problems provide a natural introduction to multigrid methods. As an example, consider the two-point boundary value problem that describes the steady-state temperature distribution in a long uniform rod. It is given by the second-order boundary value problem

$$-u''(x) + \sigma u(x) = f(x), \quad 0 < x < 1, \quad \sigma \geq 0, \tag{1.1}$$
$$u(0) = u(1) = 0. \tag{1.2}$$

While this problem can be handled analytically, our present aim is to consider numerical methods. Many such approaches are possible, the simplest of which is a finite difference method (finite element formulations will be considered in Chapter 10). The domain of the problem $\{x : 0 \leq x \leq 1\}$ is partitioned into n subintervals by introducing the grid points $x_j = jh$, where $h = 1/n$ is the constant width of the subintervals. This establishes the grid shown in Fig. 1.1, which we denote Ω^h.

At each of the $n-1$ interior grid points, the original differential equation (1.1) is replaced by a second-order finite difference approximation. In making this replacement, we also introduce v_j as an approximation to the exact solution $u(x_j)$. This approximate solution may now be represented by a vector $\mathbf{v} = (v_1, \ldots, v_{n-1})^T$, whose components satisfy the $n-1$ linear equations

$$\frac{-v_{j-1} + 2v_j - v_{j+1}}{h^2} + \sigma v_j = f(x_j), \quad 1 \leq j \leq n-1, \tag{1.3}$$
$$v_0 = v_n = 0.$$

Defining $\mathbf{f} = (f(x_1), \ldots, f(x_{n-1}))^T = (f_1, \ldots, f_{n-1})^T$, the vector of right-side values, we may also represent this system of linear equations in matrix form as

$$\frac{1}{h^2} \begin{bmatrix} 2+\sigma h^2 & -1 & & & \\ -1 & 2+\sigma h^2 & -1 & & \\ & \ddots & \ddots & \ddots & \\ & & \ddots & \ddots & -1 \\ & & & -1 & 2+\sigma h^2 \end{bmatrix} \begin{bmatrix} v_1 \\ \cdot \\ \cdot \\ \cdot \\ v_{n-1} \end{bmatrix} = \begin{bmatrix} f_1 \\ \cdot \\ \cdot \\ \cdot \\ f_{n-1} \end{bmatrix}$$

1

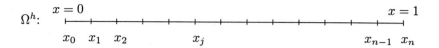

Figure 1.1: *One-dimensional grid on the interval $0 \le x \le 1$. The grid spacing is $h = \frac{1}{n}$ and the jth grid point is $x_j = jh$ for $0 \le j \le n$.*

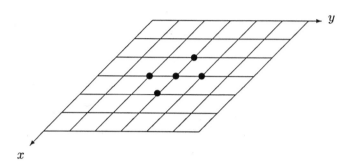

Figure 1.2: *Two-dimensional grid on the unit square. The solid dots indicate the unknowns that are related at a typical grid point by the discrete equations (1.5).*

or even more compactly as $A\mathbf{v} = \mathbf{f}$. The matrix A is $(n-1) \times (n-1)$, tridiagonal, symmetric, and positive definite.

Analogously, it is possible to formulate a two-dimensional version of this problem. Consider the second-order partial differential equation (PDE)

$$-u_{xx} - u_{yy} + \sigma u = f(x, y), \quad 0 < x < 1, \quad 0 < y < 1, \quad \sigma > 0. \tag{1.4}$$

With $\sigma = 0$, this is the Poisson equation; with $\sigma \ne 0$, it is the Helmholtz equation. We consider this equation subject to the condition that $u = 0$ on the boundary of the unit square.

As before, this problem may be cast in a discrete form by defining the grid points $(x_i, y_i) = (ih_x, jh_y)$, where $h_x = \frac{1}{m}$ and $h_y = \frac{1}{n}$. This two-dimensional grid is also denoted Ω^h and is shown in Fig. 1.2. Replacing the derivatives of (1.4) by second-order finite differences leads to the system of linear equations

$$\frac{-v_{i-1,j} + 2v_{ij} - v_{i+1,j}}{h_x^2} + \frac{-v_{i,j-1} + 2v_{ij} - v_{i,j+1}}{h_y^2} + \sigma v_{ij} = f_{ij},$$

$$\tag{1.5}$$

$$v_{i0} = v_{in} = v_{0j} = v_{mj} = 0, \quad 1 \le i \le m-1, \ 1 \le j \le n-1.$$

As before, v_{ij} is an approximation to the exact solution $u(x_i, y_j)$ and $f_{ij} = f(x_i, y_j)$.

There are now $(m-1)(n-1)$ interior grid points and the same number of unknowns in the problem. We can choose from many different orderings of the unknowns. For the moment, consider the *lexicographic* ordering by lines of constant i. The unknowns of the ith row of the grid may be collected in the vector $\mathbf{v}_i =$

$(v_{i1}, \ldots, v_{i,n-1})^T$ for $1 \leq i \leq m-1$. Similarly, let $\mathbf{f}_i = (f_{i1}, \ldots, f_{i,n-1})^T$. The system of equations (1.5) may then be given in block matrix form as

$$
\begin{bmatrix}
B & -aI & & & \\
-aI & B & -aI & & \\
& \cdot & \cdot & \cdot & \\
& & \cdot & \cdot & -aI \\
& & & -aI & B
\end{bmatrix}
\begin{bmatrix}
\mathbf{v}_1 \\ \cdot \\ \cdot \\ \cdot \\ \mathbf{v}_{m-1}
\end{bmatrix}
=
\begin{bmatrix}
\mathbf{f}_1 \\ \cdot \\ \cdot \\ \cdot \\ \mathbf{f}_{m-1}
\end{bmatrix}.
$$

This system is symmetric, block tridiagonal, and sparse. It has block dimension $(m-1) \times (m-1)$. Each diagonal block, B, is an $(n-1) \times (n-1)$ tridiagonal matrix that looks much like the matrix for the one-dimensional problem. Each off-diagonal block is a multiple, $a = \frac{1}{h_x^2}$, of the $(n-1) \times (n-1)$ identity matrix I.

Matrix Properties. The matrices produced by the discretization of self-adjoint boundary value problems have some special properties that are desirable for many numerical methods. Let A with elements a_{ij} be such a matrix. It is generally symmetric $(A = A^T)$ and sparse (a large percentage of the elements are zero). These matrices are also often *weakly diagonally dominant*, which means that, in magnitude, the diagonal element is at least as large as the sum of the off-diagonal elements in the same row:

$$
\sum_{\substack{j \neq i}}^{n} |a_{ij}| \leq |a_{ii}| \quad \text{for} \quad 1 \leq i \leq n.
$$

These matrices are also *positive definite*, which means that, for all vectors $\mathbf{u} \neq \mathbf{0}$, we have $\mathbf{u}^T A \mathbf{u} > 0$. This property is difficult to interpret, but there are several alternate characterizations. For example, a symmetric positive definite matrix has real and positive eigenvalues. It can also be shown that if A is symmetric and diagonally dominant with positive diagonal elements, then A is positive definite. One other matrix property arises in the course of our work: a symmetric positive definite matrix with positive entries on the diagonal and nonpositive off-diagonal entries is called an *M-matrix*.

We occasionally appeal to stencils associated with discrete equations. For the one-dimensional model problem, the stencil representation of the matrix is

$$
A = \frac{1}{h^2}(-1 \quad 2 + \sigma h^2 \quad -1).
$$

The two-dimensional stencil for $h_x = h_y = h$ is

$$
A = \frac{1}{h^2}
\begin{pmatrix}
& -1 & \\
-1 & 4 + \sigma h^2 & -1 \\
& -1 &
\end{pmatrix}.
$$

Stencils are useful for representing operators that interact locally on a grid. However, they must be used with care near boundaries.

The two model linear systems (1.3) and (1.5) provide the testing ground for many of the methods discussed in the following chapters. Before we proceed, however, it is useful to give a brief summary of existing methods for solving such systems.

During the past 50 years, a tremendous amount of work was devoted to the numerical solution of sparse systems of linear equations. Much of this attention was given to structured systems such as (1.3) and (1.5) that arise from boundary value problems. Existing methods of solution fall into two large categories: *direct methods* and *iterative* (or *relaxation*) methods. This tutorial is devoted to the latter category.

Direct methods, of which Gaussian elimination is the prototype, determine a solution exactly (up to machine precision) in a finite number of arithmetic steps. For systems such as (1.5) that arise from a two-dimensional elliptic equation, very efficient direct methods have been developed. They are usually based on the fast Fourier transform or the method of cyclic reduction. When applied to problems on an $n \times n$ grid, these methods require $O(n^2 \log n)$ arithmetic operations. Because they approach the minimum operation count of $O(n^2)$ operations, these methods are nearly optimal. However, they are also rather specialized and restricted primarily to systems that arise from separable self-adjoint boundary value problems.

Relaxation methods, as represented by the Jacobi and Gauss–Seidel iterations, begin with an initial guess at a solution. Their goal is to improve the current approximation through a succession of simple updating steps or iterations. The sequence of approximations that is generated (ideally) converges to the exact solution of the linear system. Classical relaxation methods are easy to implement and may be successfully applied to more general linear systems than most direct methods [23, 24, 26].

As we see in the next chapter, relaxation schemes suffer from some disabling limitations. Multigrid methods evolved from attempts to overcome these limitations. These attempts have been largely successful: used in a multigrid setting, relaxation methods are competitive with the fast direct methods when applied to the model problems, and they have more generality and a wider range of application.

In Chapters 1–5 of this tutorial, we focus on the two model problems. In Chapters 6–10, we extend the basic multigrid methods to treat more general boundary conditions, operators, and geometries. The basic methods can be applied to many elliptic and other types of problems without significant modification. Still more problems can be treated with more sophisticated multigrid methods.

Finally, the original multigrid ideas have been extended to what are more appropriately called *multilevel methods*. Purely algebraic problems (for example, network and structural problems) have led to the development of *algebraic multigrid* or *AMG*, which is the subject of Chapter 8. Beyond the boundaries of this book, multilevel methods have been applied to time-dependent problems and problems in image processing, control theory, combinatorial optimization (the traveling salesman problem), statistical mechanics (the Ising model), and quantum electrodynamics. The list of problems amenable to multilevel methods is long and growing. But first we must begin with the basics.

Exercises

1. **Derivative (Neumann) boundary conditions.** Consider model problem (1.1) subject to the *Neumann boundary conditions* $u'(0) = u'(1) = 0$. Find the system of linear equations that results when second-order finite differences are used to discretize this problem at the grid points $x_0, \ldots, x_n$. At the end

points, x_0 and x_n, one of many ways to incorporate the boundary conditions is to let $v_1 = v_0$ and $v_{n-1} = v_n$. (We return to this problem in Chapter 7.) How many equations and how many unknowns are there in this problem? Give the matrix that corresponds to this boundary value problem.

2. **Ordering unknowns.** Suppose the unknowns of system (1.5) are ordered by lines of constant j (or y). Give the block structure of the resulting matrix and specify the dimensions of the blocks.

3. **Periodic boundary conditions.** Consider model problem (1.1) subject to the *periodic boundary conditions* $u(0) = u(1)$ and $u'(0) = u'(1)$. Find the system of linear equations that results when second-order finite differences are used to discretize this problem at the grid points $x_0, \ldots, x_{n-1}$. How many equations and unknowns are there in this problem?

4. **Convection terms in two dimensions.** A convection term can be added to the two-dimensional model problem in the form

$$-\epsilon(u_{xx} + u_{yy}) + au_x = f(x).$$

Using the grid described in the text and second-order central finite difference approximations, find the system of linear equations associated with this problem. What conditions must be met by a and ϵ for the associated matrix to be diagonally dominant?

5. **Three-dimensional problem.** Consider the three-dimensional Poisson equation

$$-u_{xx} - u_{yy} - u_{zz} = f(x, y, z).$$

Write out the discrete equation obtained by using second-order central finite difference approximations at the grid point (x_i, y_j, z_k). Assuming that the unknowns are ordered first by lines of constant x, then lines of constant y, describe the block structure of the resulting matrix.

Chapter 2

Basic Iterative Methods

We now consider how model problems (1.3) and (1.5) might be treated using conventional iterative or relaxation methods. We first establish the notation for this and all remaining chapters. Let

$$A\mathbf{u} = \mathbf{f}$$

denote a system of linear equations such as (1.3) or (1.5). We always use $\mathbf{u}$ to denote the exact solution of this system and $\mathbf{v}$ to denote an approximation to the exact solution, perhaps generated by some iterative method. Bold symbols, such as $\mathbf{u}$ and $\mathbf{v}$, represent vectors, while the jth components of these vectors are denoted by u_j and v_j. In later chapters, we need to associate $\mathbf{u}$ and $\mathbf{v}$ with a particular grid, say Ω^h. In this case, the notation $\mathbf{u}^h$ and $\mathbf{v}^h$ is used.

Suppose that the system $A\mathbf{u} = \mathbf{f}$ has a unique solution and that $\mathbf{v}$ is a computed approximation to $\mathbf{u}$. There are two important measures of $\mathbf{v}$ as an approximation to $\mathbf{u}$. One is the *error* (or *algebraic error*) and is given simply by

$$\mathbf{e} = \mathbf{u} - \mathbf{v}.$$

The error is also a vector and its magnitude may be measured by any of the standard vector norms. The most commonly used norms for this purpose are the maximum (or infinity) norm and the Euclidean or 2-norm, defined, respectively, by

$$\|\mathbf{e}\|_\infty = \max_{1 \le j \le n} |e_j| \quad \text{and} \quad \|\mathbf{e}\|_2 = \left\{ \sum_{j=1}^{n} e_j^2 \right\}^{1/2}.$$

Unfortunately, the error is just as inaccessible as the exact solution itself. However, a computable measure of how well $\mathbf{v}$ approximates $\mathbf{u}$ is the *residual*, given by

$$\cdot \mathbf{r} = \mathbf{f} - A\mathbf{v}.$$

The residual is simply the amount by which the approximation $\mathbf{v}$ fails to satisfy the original problem $A\mathbf{u} = \mathbf{f}$. It is also a vector and its size may be measured by the same norm used for the error. By the uniqueness of the solution, $\mathbf{r} = \mathbf{0}$ if and only if $\mathbf{e} = \mathbf{0}$. However, it may *not* be true that when $\mathbf{r}$ is small in norm, $\mathbf{e}$ is also small in norm.

Residuals and Errors. A residual may be defined for any numerical approximation and, in many cases, a small residual does *not* necessarily imply a small error. This is certainly true for systems of linear equations, as shown by the following two problems:

$$\begin{pmatrix} 1 & -1 \\ 21 & -20 \end{pmatrix} \begin{pmatrix} u_1 \\ u_2 \end{pmatrix} = \begin{pmatrix} -1 \\ -19 \end{pmatrix} \quad \text{and} \quad \begin{pmatrix} 1 & -1 \\ 3 & -1 \end{pmatrix} \begin{pmatrix} u_1 \\ u_2 \end{pmatrix} = \begin{pmatrix} -1 \\ 1 \end{pmatrix}.$$

Both systems have the exact solution $\mathbf{u} = (1,2)^T$. Suppose we have computed the approximation $\mathbf{v} = (1.95, 3)^T$. The error in this approximation is $\mathbf{e} = (-0.95, -1)^T$, for which $\|\mathbf{e}\|_2 = 1.379$. The norm of the residual in $\mathbf{v}$ for the first system is $\|\mathbf{r}_1\|_2 = 0.071$, while the residual norm for the second system is $\|\mathbf{r}_2\|_2 = 1.851$. Clearly, the relatively small residual for the first system does not reflect the rather large error. See Exercise 18 for an important relationship between error and residual norms.

Remembering that $A\mathbf{u} = \mathbf{f}$ and using the definitions of $\mathbf{r}$ and $\mathbf{e}$, we can derive an extremely important relationship between the error and the residual (Exercise 2):

$$A\mathbf{e} = \mathbf{r}.$$

We call this relationship the *residual equation*. It says that the error satisfies the same set of equations as the unknown $\mathbf{u}$ when $\mathbf{f}$ is replaced by the residual $\mathbf{r}$. The residual equation plays a vital role in multigrid methods and it is used repeatedly throughout this tutorial.

We can now anticipate, in an imprecise way, how the residual equation can be used to great advantage. Suppose that an approximation $\mathbf{v}$ has been computed by some method. It is easy to compute the residual $\mathbf{r} = \mathbf{f} - A\mathbf{v}$. To improve the approximation $\mathbf{v}$, we might solve the residual equation for $\mathbf{e}$ and then compute a new approximation using the definition of the error

$$\mathbf{u} = \mathbf{v} + \mathbf{e}.$$

In practice, this method must be applied more carefully than we have indicated. Nevertheless, this idea of residual correction is very important in all that follows.

We now turn to relaxation methods for our first model problem (1.3) with $\sigma = 0$. Multiplying that equation by h^2 for convenience, the discrete problem becomes

$$\begin{aligned} -u_{j-1} + 2u_j - u_{j+1} &= h^2 f_j, \qquad 1 \leq j \leq n-1, \\ u_0 = u_n &= 0. \end{aligned} \tag{2.1}$$

One of the simplest schemes is the *Jacobi* (or simultaneous displacement) method. It is produced by solving the jth equation of (2.1) for the jth unknown and using the current approximation for the $(j-1)$st and $(j+1)$st unknowns. Applied to the vector of current approximations, this produces an iteration scheme that may be written in component form as

$$v_j^{(1)} = \frac{1}{2}\left(v_{j-1}^{(0)} + v_{j+1}^{(0)} + h^2 f_j\right), \qquad 1 \leq j \leq n-1.$$

To keep the notation as simple as possible, the current approximation (or the initial guess on the first iteration) is denoted $\mathbf{v}^{(0)}$, while the new, updated approximation is denoted $\mathbf{v}^{(1)}$. In practice, once all of the $\mathbf{v}^{(1)}$ components have been computed, the procedure is repeated, with $\mathbf{v}^{(1)}$ playing the role of $\mathbf{v}^{(0)}$. These iteration sweeps are continued until (ideally) convergence to the solution is obtained.

It is important to express these relaxation schemes in matrix form, as well as component form. We split the matrix A in the form

$$A = D - L - U,$$

where D is the diagonal of A, and $-L$ and $-U$ are the strictly lower and upper triangular parts of A, respectively. Including the h^2 term in the vector $\mathbf{f}$, then $A\mathbf{u} = \mathbf{f}$ becomes

$$(D - L - U)\mathbf{u} = \mathbf{f}.$$

Isolating the diagonal terms of A, we have

$$D\mathbf{u} = (L + U)\mathbf{u} + \mathbf{f}$$

or

$$\mathbf{u} = D^{-1}(L + U)\mathbf{u} + D^{-1}\mathbf{f}.$$

Multiplying by D^{-1} corresponds exactly to solving the jth equation for u_j, for $1 \leq j \leq n - 1$. If we define the Jacobi iteration matrix by

$$R_J = D^{-1}(L + U),$$

then the Jacobi method appears in matrix form as

$$\mathbf{v}^{(1)} = R_J \mathbf{v}^{(0)} + D^{-1}\mathbf{f}.$$

There is a simple but important modification that can be made to the Jacobi iteration. As before, we compute the new Jacobi iterates using

$$v_j^* = \frac{1}{2}\left(v_{j-1}^{(0)} + v_{j+1}^{(0)} + h^2 f_j\right), \qquad 1 \leq j \leq n - 1.$$

However, v_j^* is now only an intermediate value. The new iterate is given by the weighted average

$$v_j^{(1)} = (1 - \omega)v_j^{(0)} + \omega v_j^* = v_j^{(0)} + \omega(v_j^* - v_j^{(0)}), \qquad 1 \leq j \leq n - 1,$$

where $\omega \in \mathbf{R}$ is a weighting factor that may be chosen. This generates an entire family of iterations called the *weighted* or *damped Jacobi* method. Notice that $\omega = 1$ yields the original Jacobi iteration.

In matrix form, the weighted Jacobi method is given by (Exercise 3)

$$\mathbf{v}^{(1)} = [(1 - \omega)I + \omega R_J]\mathbf{v}^{(0)} + \omega D^{-1}\mathbf{f}.$$

If we define the weighted Jacobi iteration matrix by

$$R_\omega = (1 - \omega)I + \omega R_J,$$

then the method may be expressed as (Exercise 3)

$$\mathbf{v}^{(1)} = R_\omega \mathbf{v}^{(0)} + \omega D^{-1} \mathbf{f}.$$

We should note in passing that the weighted Jacobi iteration can also be written in the form (Exercise 3)

$$\mathbf{v}^{(1)} = \mathbf{v}^{(0)} + \omega D^{-1} \mathbf{r}^{(0)}.$$

This says that the new approximation is obtained from the current one by adding an appropriate weighting of the residual.

This is just one example of a *stationary linear iteration*. This term refers to the fact that the update rule is linear in the unknown $\mathbf{v}$ and does not change from one iteration to the next. We can say more about such iterations in general. Recalling that $\mathbf{e} = \mathbf{u} - \mathbf{v}$ and $A\mathbf{e} = \mathbf{r}$, we have

$$\mathbf{u} - \mathbf{v} = A^{-1} \mathbf{r}.$$

Identifying $\mathbf{v}$ with the current approximation $\mathbf{v}^{(0)}$ and $\mathbf{u}$ with the new approximation $\mathbf{v}^{(1)}$, an iteration may be formed by taking

$$\mathbf{v}^{(1)} = \mathbf{v}^{(0)} + B\mathbf{r}^{(0)}, \tag{2.2}$$

where B is an approximation to A^{-1}. If B can be chosen "close" to A^{-1}, then the iteration should be effective.

It is useful to examine this general form of iteration a bit further. Rewriting expression (2.2), we see that

$$\begin{aligned}
\mathbf{v}^{(1)} = \mathbf{v}^{(0)} + B\mathbf{r}^{(0)} &= \mathbf{v}^{(0)} + B(f - A\mathbf{v}^{(0)}) \\
&= (I - BA)\mathbf{v}^{(0)} + Bf \\
&\equiv R\mathbf{v}^{(0)} + Bf,
\end{aligned}$$

where we have defined the general iteration matrix as $R = I - BA$. It can also be shown (Exercise 4) that m sweeps of this iteration result in

$$\mathbf{v}^{(m)} = R^m \mathbf{v}^{(0)} + C(\mathbf{f}),$$

where $C(\mathbf{f})$ represents a series of operations on $\mathbf{f}$. We return to this general form in Chapter 5.

Before analyzing or implementing these methods, we present a few more of the basic iterative schemes. Weighted Jacobi computes all components of the new approximation before using any of them. This requires $2n$ storage locations for the approximation vector. It also means that new information cannot be used as soon as it is available.

The *Gauss–Seidel* method incorporates a simple change: components of the new approximation are used as soon as they are computed. This means that components of the approximation vector $\mathbf{v}$ are overwritten as soon as they are updated. This small change reduces the storage requirement for the approximation vector to only n locations. The Gauss–Seidel method is also equivalent to successively setting each component of the residual vector to zero and solving for the corresponding component of the solution (Exercise 5). When applied to the model problem, this method may be expressed in component form as

$$v_j \longleftarrow \frac{1}{2}\left(v_{j-1} + v_{j+1} + h^2 f_j\right), \qquad 1 \le j \le n-1,$$

where the arrow notation stands for replacement or overwriting.

Once again it is useful to express this method in matrix form. Splitting the matrix A in the form $A = D - L - U$, we can now write the original system of equations as

$$(D - L)\mathbf{u} = U\mathbf{u} + \mathbf{f}$$

or

$$\mathbf{u} = (D - L)^{-1}U\mathbf{u} + (D - L)^{-1}\mathbf{f}.$$

This representation corresponds to solving the jth equation for u_j and using new approximations for components $1, 2, \ldots, j - 1$. Defining the Gauss–Seidel iteration matrix by

$$R_G = (D - L)^{-1}U,$$

we can express the method as

$$\mathbf{v} \longleftarrow R_G\mathbf{v} + (D - L)^{-1}\mathbf{f}.$$

Finally, we look at one important variation on the Gauss–Seidel iteration. For weighted Jacobi, the order in which the components of $\mathbf{v}$ are updated is immaterial, since components are never overwritten. However, for Gauss–Seidel, the order of updating is significant. Instead of sweeping through the components (equivalently, the grid points) in ascending order, we could sweep through the components in descending order or we might alternate between ascending and descending orders. The latter procedure is called the *symmetric Gauss–Seidel* method.

Another effective alternative is to update all the even components first by the expression

$$v_{2j} \longleftarrow \frac{1}{2}\left(v_{2j-1} + v_{2j+1} + h^2 f_{2j}\right),$$

and then update all the odd components using

$$v_{2j+1} \longleftarrow \frac{1}{2}\left(v_{2j} + v_{2j+2} + h^2 f_{2j+1}\right).$$

This strategy leads to the *red-black Gauss–Seidel* method, which is illustrated in Fig. 2.1 for both one-dimensional and two-dimensional grids. Notice that the red points correspond to even-indexed points in one dimension and to points whose index sum is even in two dimensions (assuming that $i = 0$ and $j = 0$ corresponds to a boundary). The red points also correspond to what we soon call coarse-grid points.

The advantages of red-black over regular Gauss–Seidel are not immediately apparent; the issue is often problem-dependent. However, red-black Gauss–Seidel does have a clear advantage in terms of parallel computation. The red points need only the black points for their updating and may therefore be updated in any order. This work represents $\frac{n}{2}$ (or $\frac{n^2}{2}$ in two dimensions) independent tasks that can be distributed among several independent processors. In a similar way, the black sweep can also be done by several independent processors. (The Jacobi iteration is also well-suited to parallel computation.)

There are many more basic iterative methods. However, we have seen enough of the essential methods to move ahead toward multigrid. First, it is important to gain some understanding of how these basic iterations perform. We proceed both by analysis and by experimentation.

When studying stationary linear iterations, it is sufficient to work with the homogeneous linear system $A\mathbf{u} = \mathbf{0}$ and use arbitrary initial guesses to start the

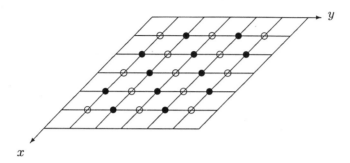

Figure 2.1: *A one-dimensional grid (top) and a two-dimensional grid (bottom), showing the red points (○) and the black points (●) for red-black relaxation.*

relaxation scheme (Exercise 6). One reason for doing this is that the exact solution is known ($\mathbf{u} = \mathbf{0}$) and the error in an approximation $\mathbf{v}$ is simply $-\mathbf{v}$. Therefore, we return to the one-dimensional model problem with $\mathbf{f} = \mathbf{0}$. It appears as

$$
\begin{aligned}
-u_{j-1} + 2u_j - u_{j+1} &= 0, \qquad 1 \le j \le n - 1, \\
u_0 = u_n &= 0.
\end{aligned}
\tag{2.3}
$$

We obtain some valuable insight by applying various iterations to this system of equations with an initial guess consisting of the vectors (or *Fourier modes*)

$$
v_j = \sin\left(\frac{jk\pi}{n}\right), \qquad 0 \le j \le n, \ 1 \le k \le n - 1.
$$

Recall that j denotes the component (or associated grid point) of the vector $\mathbf{v}$. The integer k now makes its first appearance. It is called the *wavenumber* (or *frequency*) and it indicates the number of half sine waves that constitute $\mathbf{v}$ on the domain of the problem. We use $\mathbf{v}_k$ to designate the entire vector $\mathbf{v}$ with wavenumber k. Figure 2.2 illustrates initial guesses $\mathbf{v}_1$, $\mathbf{v}_3$, and $\mathbf{v}_6$. Notice that small values of k correspond to long, smooth waves, while large values of k correspond to highly oscillatory waves. We now explore how Fourier modes behave under iteration.

We first apply the weighted Jacobi iteration with $\omega = \frac{2}{3}$ to problem (2.3) on a grid with $n = 64$ points. Beginning with initial guesses of $\mathbf{v}_1$, $\mathbf{v}_3$, and $\mathbf{v}_6$, the iteration is applied 100 times. Recall that the error is just $-\mathbf{v}$. Figure 2.3(a) shows a plot of the maximum norm of the error versus the iteration number.

For the moment, only the qualitative behavior of the iteration is important. The error clearly decreases with each relaxation sweep and the rate of decrease is larger for the higher wavenumbers. Figures 2.3(b, c) show analogous plots for the regular and red-black Gauss–Seidel iterations, where we see a similar relationship among the error, the number of iterations, and the wavenumber. (The complete situation is not quite so simple with red-black Gauss–Seidel, as illustrated in Exercise 20.)

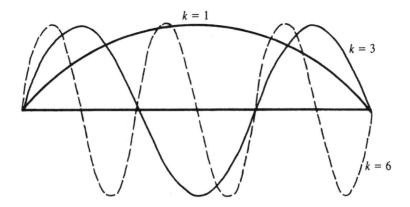

Figure 2.2: *The modes* $v_j = \sin\left(\frac{jk\pi}{n}\right)$, $0 \le j \le n$, *with wavenumbers* $k = 1, 3, 6$. *The kth mode consists of* $\frac{k}{2}$ *full sine waves on the interval.*

The experiment of Fig. 2.3(a) is presented in a slightly different light in Fig. 2.4. In this figure, the log of the maximum norm of the error for the weighted Jacobi method is plotted against the iteration number for various wavenumbers. This plot clearly shows a linear decrease in the log of the error norm, indicating that the error itself decreases geometrically with each iteration. If we let $\mathbf{e}^{(0)}$ be the error in the initial guess and $\mathbf{e}^{(m)}$ be the error in the mth iterate, then we might expect to describe the error by a relationship of the form

$$\|\mathbf{e}^{(m)}\|_\infty = c_k^m \|\mathbf{e}^{(0)}\|_\infty,$$

where c_k is a constant that depends on the wavenumber. We will see that the theory confirms this conjecture.

In general, most initial guesses (or, equivalently, most right-side vectors $\mathbf{f}$) would not consist of a single mode. Consider a slightly more realistic situation in which the initial guess (hence, the error) consists of three modes: a low-frequency wave ($k = 1$), a medium-frequency wave ($k = 6$), and a high-frequency wave ($k = 32$) on a grid with $n = 64$ points; it is given by

$$v_j = \frac{1}{3}\left[\sin\left(\frac{j\pi}{n}\right) + \sin\left(\frac{6j\pi}{n}\right) + \sin\left(\frac{32j\pi}{n}\right)\right].$$

Figure 2.5 shows the maximum norm of the error plotted against the number of iterations. The error decreases rapidly within the first five iterations, after which it decreases much more slowly. The initial decrease corresponds to the quick elimination of the high-frequency modes of the error. The slow decrease is due to the presence of persistent low-frequency modes. The important observation is that the standard iterations converge very quickly as long as the error has high-frequency components. However, the slower elimination of the low-frequency components degrades the performance of these methods.

With some experimental evidence in hand, we now turn to a more analytical approach. Each of the methods discussed so far may be represented in the form

$$\mathbf{v}^{(1)} = R\mathbf{v}^{(0)} + \mathbf{g},$$

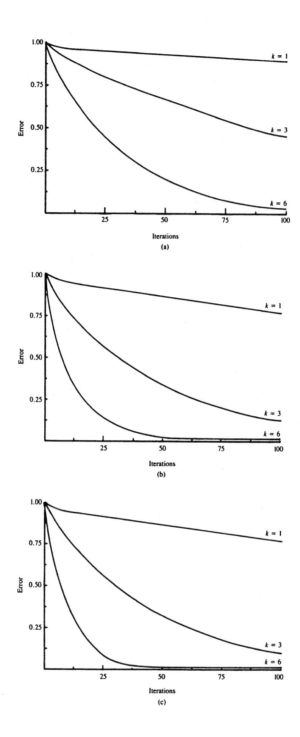

Figure 2.3: (a) *Weighted Jacobi iteration with* $\omega = \frac{2}{3}$, (b) *regular Gauss–Seidel iteration, and* (c) *red-black Gauss–Seidel iteration applied to the one-dimensional model problem with* $n = 64$ *points and with initial guesses* $\mathbf{v}_1$, $\mathbf{v}_3$, *and* $\mathbf{v}_6$. *The maximum norm of the error,* $\|\mathbf{e}\|_\infty$, *is plotted against the iteration number for* 100 *iterations.*

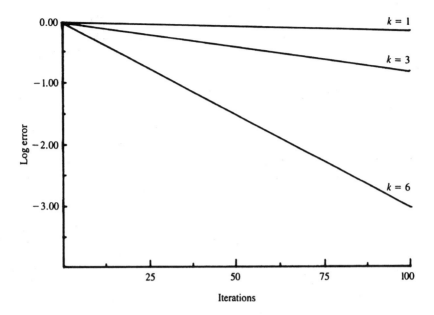

Figure 2.4: *Weighted Jacobi iteration with $\omega = \frac{2}{3}$ applied to the one-dimensional model problem with $n = 64$ points and with initial guesses $\mathbf{v}_1$, $\mathbf{v}_3$, and $\mathbf{v}_6$. The log of $\|\mathbf{e}\|_\infty$ is plotted against the iteration number for 100 iterations.*

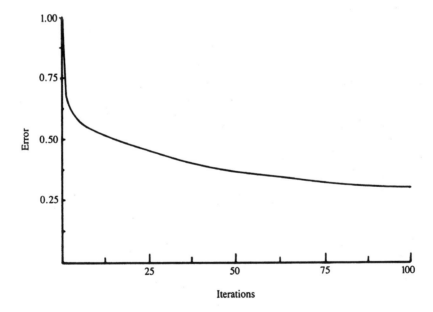

Figure 2.5: *Weighted Jacobi method with $\omega = \frac{2}{3}$ applied to the one-dimensional model problem with $n = 64$ points and an initial guess $(\mathbf{v}_1 + \mathbf{v}_6 + \mathbf{v}_{32})/3$. The maximum norm of the error, $\|\mathbf{e}\|_\infty$, is plotted against the iteration number for 100 iterations.*

where R is one of the iteration matrices derived earlier. Furthermore, all of these methods are designed such that the exact solution, $\mathbf{u}$, is a fixed point of the iteration (Exercise 4). This means that iteration does not change the exact solution:

$$\mathbf{u} = R\mathbf{u} + \mathbf{g}.$$

Subtracting these last two expressions, we find that

$$\mathbf{e}^{(1)} = R\mathbf{e}^{(0)}.$$

Repeating this argument, it follows that after m relaxation sweeps, the error in the mth approximation is given by

$$\mathbf{e}^{(m)} = R^m \mathbf{e}^{(0)}.$$

Matrix Norms. Matrix norms can be defined in terms of the commonly used vector norms. Let A be an $n \times n$ matrix with elements a_{ij}. Consider the vector norm $\|\mathbf{x}\|_p$ defined by

$$\|\mathbf{x}\|_p = \left(\sum_{i=1}^n |x_i|^p \right)^{1/p}, \quad 1 \leq p < \infty,$$

$$\|\mathbf{x}\|_\infty = \sup_{1 \leq i \leq n} |x_i|.$$

The matrix norm *induced* by the vector norm $\| \cdot \|_p$ is defined by

$$\|A\|_p = \sup_{\mathbf{x} \neq 0} \frac{\|A\mathbf{x}\|_p}{\|\mathbf{x}\|_p}.$$

While not obvious without some computation, this definition leads to the following matrix norms induced by the vector norms $\| \cdot \|_1$, $\| \cdot \|_\infty$, and $\| \cdot \|_2$:

$\|A\|_1 = \max_j \sum_{i=1}^n |a_{ij}|$ (maximum column sum),

$\|A\|_\infty = \max_i \sum_{j=1}^n |a_{ij}|$ (maximum row sum),

$\|A\|_2 = \sqrt{\text{spectral radius of } A^T A}$.

Recall that the *spectral radius* of a matrix is given by

$$\rho(A) = \max |\lambda(A)|,$$

where $\lambda(A)$ denotes the eigenvalues of A. For symmetric matrices, the matrix 2-norm is just the spectral radius of A:

$$\|A\|_2 = \sqrt{\rho(A^T A)} = \sqrt{\rho(A^2)} = \rho(A).$$

If we now choose a particular vector norm and its associated matrix norm, it is possible to bound the error after m iterations by

$$\|\mathbf{e}^{(m)}\| \leq \|R\|^m \|\mathbf{e}^{(0)}\|.$$

This leads us to conclude that if $\|R\| < 1$, then the error is forced to zero as the iteration proceeds.

It is shown in many standard texts [9, 20, 24, 26] that

$$\lim_{m \to \infty} R^m = 0 \quad \text{if and only if} \quad \rho(R) < 1.$$

(Therefore, it follows that the iteration associated with the matrix R converges for all initial guesses if and only if $\rho(R) < 1$.) *Convergence is achieved when $\rho(R)$*

The spectral radius $\rho(R)$ is also called the *asymptotic convergence factor* when *is computed* it appears in the context of iterative methods. It has some useful interpretations. *and* First, it is roughly the worst factor by which the error is reduced with each relax- *shown < 1* ation sweep. By the following argument, it also tells us approximately how many iterations are required to reduce the error by a factor of 10^{-d}. Let m be the smallest integer that satisfies

$$\frac{\|\mathbf{e}^{(m)}\|}{\|\mathbf{e}^{(0)}\|} \le 10^{-d}.$$

$1 - \frac{\omega}{2}\lambda(A) < 1$

This condition will be approximately satisfied if   *$\lambda\mathbf{e} = A\mathbf{e}$* *$\frac{1-\omega}{2}$.*

$$[\rho(R)]^m \le 10^{-d}.$$

$0 < \frac{\omega}{2}\lambda(A) < \frac{1}{2}$

Solving for m, we have

$1 < 1 - \frac{\omega}{2}\lambda(A) < 1 - \frac{1}{2}$

$$m \ge -\frac{d}{\log_{10}[\rho(R)]}.$$

$1 < 1 - \frac{\omega}{2}\lambda(A) < \frac{1}{2}$

The quantity $-\log_{10}(\rho(R))$ is called the *asymptotic convergence rate*. Its reciprocal gives the approximate number of iterations required to reduce the error by one decimal digit. We see that as $\rho(R)$ approaches 1, the convergence rate decreases. Small values of $\rho(R)$ (that is, $\rho(R)$ positive and near zero) give a high convergence rate.

We have established the importance of the spectral radius of the iteration matrix in analyzing the convergence properties of relaxation methods. Now it is time to compute some spectral radii. Consider the weighted Jacobi iteration applied to the one-dimensional model problem. Recalling that $R_\omega = (1 - \omega)I + \omega R_J$, we have (Exercise 3)

$$R_\omega = I - \frac{\omega}{2}\begin{bmatrix} 2 & -1 & & & \\ -1 & 2 & -1 & & \\ & \cdot & \cdot & \cdot & \\ & & \cdot & \cdot & \cdot \\ & & & \cdot & \cdot & -1 \\ & & & & -1 & 2 \end{bmatrix}.$$

Written in this form, it follows that the eigenvalues of R_ω and A are related by

$$\lambda(R_\omega) = 1 - \frac{\omega}{2}\lambda(A).$$

The problem becomes one of finding the eigenvalues of the original matrix A. This useful exercise (Exercise 8) may be done in several different ways. The result is

Interpreting the Spectral Radius. The spectral radius is considered to be an *asymptotic* measure of convergence because it predicts the worst-case error reduction over many iterations. It can be shown [9, 20] that, in any vector norm,

$$\rho(R) = \lim_{m \to \infty} \|R^m\|^{1/m}.$$

Therefore, in terms of error reduction, we have

$$\rho(R) = \lim_{m \to \infty} \sup_{\mathbf{e}^{(0)}} \left(\frac{\|\mathbf{e}^{(m)}\|}{\|\mathbf{e}^{(0)}\|} \right)^{1/m}.$$

However, the spectral radius does not, in general, predict the behavior of the error norm for a single iteration. For example, consider the matrix

$$R = \begin{pmatrix} 0 & 100 \\ 0 & 0 \end{pmatrix}.$$

Clearly, $\rho(R) = 0$. But if we start with $\mathbf{e}^{(0)} = (0, 1)^T$ and compute $\mathbf{e}^{(1)} = R\mathbf{e}^{(0)}$, then the convergence factor is

$$\frac{\|\mathbf{e}^{(1)}\|_2}{\|\mathbf{e}^{(0)}\|_2} = 100.$$

The next iterate achieves the asymptotic estimate, $\rho(R) = 0$, because $\mathbf{e}^{(2)} = 0$. A better worst-case estimate of error reduction for one or a few iterations is given by the matrix norm $\|R\|_2$. For the above example, we have $\|R\|_2 = 100$. The discrepancy between the asymptotic convergence factor, $\rho(R)$, and the worst-case estimate, $\|R\|_2$, disappears when R is symmetric because then $\rho(R) = \|R\|_2$.

that the eigenvalues of A are

$$\lambda_k(A) = 4\sin^2\left(\frac{k\pi}{2n}\right), \qquad 1 \le k \le n-1.$$

Also of interest are the corresponding eigenvectors of A. In all that follows, we let $w_{k,j}$ be the jth component of the kth eigenvector, $\mathbf{w}_k$. The eigenvectors of A are then given by (Exercise 9)

$$w_{k,j} = \sin\left(\frac{jk\pi}{n}\right), \qquad 1 \le k \le n-1, \quad 0 \le j \le n.$$

We see that the eigenvectors of A are simply the Fourier modes discussed earlier. With these results, we find that the eigenvalues of R_ω are

$$\lambda_k(R_\omega) = 1 - 2\omega\sin^2\left(\frac{k\pi}{2n}\right), \qquad 1 \le k \le n-1,$$

while the eigenvectors of R_ω are the same as the eigenvectors of A (Exercise 10). It is important to note that if $0 < \omega \le 1$, then $|\lambda_k(R_\omega)| < 1$ and the weighted Jacobi iteration converges. We return to these convergence properties in more detail after a small detour.

The eigenvectors of the matrix A are important in much of the following discussion. They correspond very closely to the eigenfunctions of the continuous model problem. Just as we can expand fairly arbitrary functions using this set of eigenfunctions, it is also possible to expand arbitrary vectors in terms of a set of eigenvectors. Let $\mathbf{e}^{(0)}$ be the error in an initial guess used in the weighted Jacobi method. Then it is possible to represent $\mathbf{e}^{(0)}$ using the eigenvectors of A in the form

$$\mathbf{e}^{(0)} = \sum_{k=1}^{n-1} c_k \mathbf{w}_k,$$

where the coefficients $c_k \in \mathbf{R}$ give the "amount" of each mode in the error. We have seen that after m sweeps of the iteration, the error is given by

$$\mathbf{e}^{(m)} = R_\omega^m \mathbf{e}^{(0)}.$$

Using the eigenvector expansion for $\mathbf{e}^{(0)}$, we have

$$\mathbf{e}^{(m)} = R_\omega^m \mathbf{e}^{(0)} = \sum_{k=1}^{n-1} c_k R_\omega^m \mathbf{w}_k = \sum_{k=1}^{n-1} c_k \lambda_k^m(R_\omega) \mathbf{w}_k.$$

The last equality follows because the eigenvectors of A and R_ω are the same; therefore, $R_\omega \mathbf{w}_k = \lambda_k(R_\omega) \mathbf{w}_k$.

This expansion for $\mathbf{e}^{(m)}$ shows that after m iterations, the kth mode of the initial error has been reduced by a factor of $\lambda_k^m(R_\omega)$. It should also be noted that the weighted Jacobi method does not mix modes: when applied to a single mode, the iteration can change the amplitude of that mode, but it cannot convert that mode into different modes. In other words, the Fourier modes are also eigenvectors of the iteration matrix. As we will see, this property is not shared by all stationary iterations.

To develop some familiarity with these Fourier modes, Fig. 2.6 shows them on a grid with $n = 12$ points. Notice that the kth mode consists of $\frac{k}{2}$ full sine waves and has a wavelength of $\ell = \frac{24h}{k} = \frac{2}{k}$ (the entire interval has length 1). The $k = \frac{n}{2}$ mode has a wavelength of $\ell = 4h$ and the $k = n - 1$ mode has a wavelength of almost $\ell = 2h$. Waves with wavenumbers greater than n (wavelengths less than $2h$) cannot be represented on the grid. In fact (Exercise 12), through the phenomenon of *aliasing*, a wave with a wavelength less than $2h$ actually appears on the grid with a wavelength greater than $2h$.

At this point, it is important to establish some terminology that is used throughout the remainder of the tutorial. We need some qualitative terms for the various Fourier modes that have been discussed. The modes in the lower half of the spectrum, with wavenumbers in the range $1 \le k < \frac{n}{2}$, are called *low-frequency* or *smooth* modes. The modes in the upper half of the spectrum, with $\frac{n}{2} \le k \le n - 1$, are called *high-frequency* or *oscillatory* modes.

Having taken this excursion through Fourier modes, we now return to the analysis of the weighted Jacobi method. We established that the eigenvalues of the iteration matrix are given by

$$\lambda_k(R_\omega) = 1 - 2\omega \sin^2\left(\frac{k\pi}{2n}\right), \qquad 1 \le k \le n - 1.$$

What choice of ω gives the best iterative scheme?

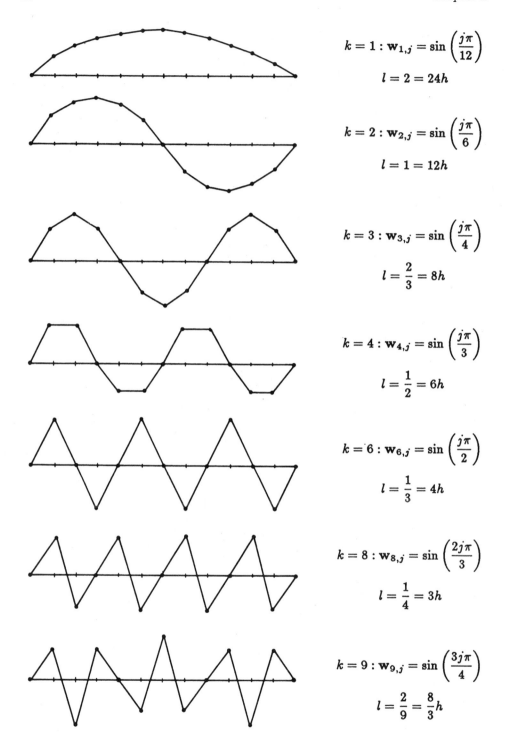

$$k = 1 : \mathbf{w}_{1,j} = \sin\left(\frac{j\pi}{12}\right)$$

$$l = 2 = 24h$$

$$k = 2 : \mathbf{w}_{2,j} = \sin\left(\frac{j\pi}{6}\right)$$

$$l = 1 = 12h$$

$$k = 3 : \mathbf{w}_{3,j} = \sin\left(\frac{j\pi}{4}\right)$$

$$l = \frac{2}{3} = 8h$$

$$k = 4 : \mathbf{w}_{4,j} = \sin\left(\frac{j\pi}{3}\right)$$

$$l = \frac{1}{2} = 6h$$

$$k = 6 : \mathbf{w}_{6,j} = \sin\left(\frac{j\pi}{2}\right)$$

$$l = \frac{1}{3} = 4h$$

$$k = 8 : \mathbf{w}_{8,j} = \sin\left(\frac{2j\pi}{3}\right)$$

$$l = \frac{1}{4} = 3h$$

$$k = 9 : \mathbf{w}_{9,j} = \sin\left(\frac{3j\pi}{4}\right)$$

$$l = \frac{2}{9} = \frac{8}{3}h$$

Figure 2.6: *Graphs of the Fourier modes of A on a grid with $n = 12$ points. Modes with wavenumbers $k = 1, 2, 3, 4, 6, 8, 9$ are shown. The wavelength of the kth mode is $\ell = \frac{24h}{k}$.*

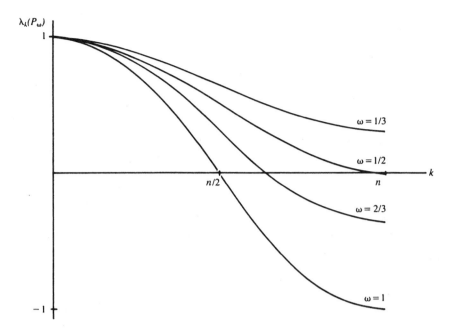

Figure 2.7: *Eigenvalues of the iteration matrix R_ω for $\omega = \frac{1}{3}, \frac{1}{2}, \frac{2}{3}, 1$. The eigenvalues $\lambda_k = 1 - 2\omega \sin^2\left(\frac{k\pi}{2n}\right)$ are plotted as if k were a continuous variable on the interval $0 \le k \le n$. In fact, $1 \le k \le n-1$ takes only integer values.*

Recall that for $0 < \omega \le 1$, we have $|\lambda_k(R_\omega)| < 1$. We would like to find the value of ω that makes $|\lambda_k(R_\omega)|$ as small as possible for all $1 \le k \le n-1$. Figure 2.7 is a plot of the eigenvalues λ_k for four different values of ω. Notice that for all values of ω satisfying $0 < \omega \le 1$,

$$\lambda_1 = 1 - 2\omega \sin^2\left(\frac{\pi}{2n}\right) = 1 - 2\omega \sin^2\left(\frac{\pi h}{2}\right) \approx 1 - \frac{\omega \pi^2 h^2}{2}.$$

This fact implies that λ_1, the eigenvalue associated with the smoothest mode, will always be close to 1. Therefore, no value of ω will reduce the smooth components of the error effectively. Furthermore, the smaller the grid spacing h, the closer λ_1 is to 1. Any attempt to improve the accuracy of the solution (by decreasing the grid spacing) will only worsen the convergence of the smooth components of the error. Most basic relaxation schemes share this ironic limitation.

Having accepted the fact that no value of ω damps the smooth components satisfactorily, we ask what value of ω provides the best damping of the oscillatory components (those with $\frac{n}{2} \le k \le n-1$). We could impose this condition by requiring that

$$\lambda_{n/2}(R_\omega) = -\lambda_n(R_\omega).$$

Solving this equation for ω leads to the optimal value $\omega = \frac{2}{3}$.

We also find (Exercise 13) that with $\omega = \frac{2}{3}$, $|\lambda_k| < \frac{1}{3}$ for all $\frac{n}{2} \le k \le n-1$. This says that the oscillatory components are reduced at least by a factor of three with each relaxation. This damping factor for the oscillatory modes is an important property of any relaxation scheme and is called the *smoothing factor* of the scheme. An important property of the basic relaxation scheme that underlies much of the power of multigrid methods is that the smoothing factor is not only small, but also independent of the grid spacing h.

We now turn to some numerical experiments to illustrate the analytical results that have just been obtained. Once again, the weighted Jacobi method is applied to the one-dimensional model problem $A\mathbf{u} = \mathbf{0}$ on a grid with $n = 64$ points. We use initial guesses (which are also initial errors) consisting of single modes with wavenumbers $1 \leq k \leq n - 1$. Figure 2.8 shows how the method performs in terms of different wavenumbers. Specifically, the wavenumber of the initial error is plotted against the number of iterations required to reduce the norm of the initial error by a factor of 100. This experiment is done for weighting factors of $\omega = 1$ and $\omega = \frac{2}{3}$.

With $\omega = 1$, both the high- and low-frequency components of the error are damped very slowly. Components with wavenumbers near $\frac{n}{2}$ are damped rapidly. This behavior is consistent with the eigenvalue curves of Fig. 2.7. We see a quite different behavior in Fig. 2.8(b) with $\omega = \frac{2}{3}$. Recall that $\omega = \frac{2}{3}$ was chosen to give preferential damping to the oscillatory components. Indeed, the smooth waves are damped very slowly, while the upper half of the spectrum ($k \geq \frac{n}{2}$) shows rapid convergence. Again, this is consistent with Fig. 2.7.

Another perspective on these convergence properties is provided in Figure 2.9. This time the actual approximations are plotted. The weighted Jacobi method with $\omega = \frac{2}{3}$ is applied to the same model problem on a grid with $n = 64$ points. Figure 2.9(a) shows the error with wavenumber $k = 3$ after one relaxation sweep (left plot) and after 10 relaxation sweeps (right plot). This smooth component is damped very slowly. Figure 2.9(b) shows a more oscillatory error ($k = 16$) after one and after 10 iterations. The damping is now much more dramatic. Notice also, as mentioned before, that the weighted Jacobi method preserves modes: once a $k = 3$ mode, always a $k = 3$ mode.

Figure 2.9(c) illustrates the selectivity of the damping property. This experiment uses an initial guess consisting of two modes with $k = 2$ and $k = 16$. After 10 relaxation sweeps, the high-frequency modulation on the long wave has been nearly eliminated. However, the original smooth component persists.

We have belabored the discussion of the weighted Jacobi method because it is easy to analyze and because it shares many properties with other basic relaxation schemes. In much less detail, let us look at the Gauss–Seidel iteration. We can show (Exercise 14) that the Gauss–Seidel iteration matrix for the model problem (matrix A) has eigenvalues

$$\lambda_k(R_G) = \cos^2\left(\frac{k\pi}{n}\right), \qquad 1 \leq k \leq n - 1.$$

These eigenvalues, which are plotted in Fig. 2.10, must be interpreted carefully. We see that when k is close to 1 or n, the corresponding eigenvalues are close to 1 and convergence is slow. However, the eigenvectors of R_G are given by (Exercise 14)

$$w_{k,j} = \left[\cos\left(\frac{k\pi}{n}\right)\right]^j \sin\left(\frac{jk\pi}{n}\right),$$

where $0 \leq j \leq n$ and $1 \leq k \leq n - 1$. These eigenvectors do not coincide with the eigenvectors of A. Therefore, $\lambda_k(R_G)$ gives the convergence rate, not for the kth mode of A, but for the kth eigenvector of R_G.

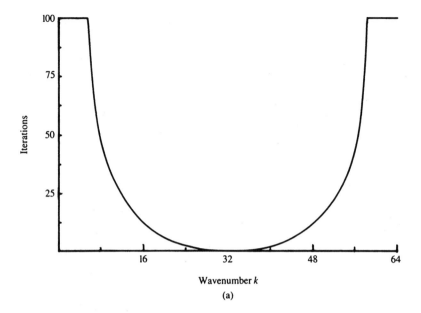

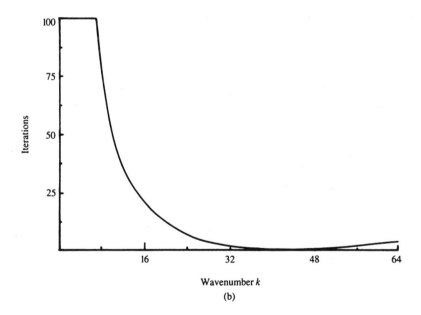

Figure 2.8: *Weighted Jacobi method with (a) $\omega = 1$ and (b) $\omega = \frac{2}{3}$ applied to the one-dimensional model problem with $n = 64$ points. The initial guesses consist of the modes $\mathbf{w}_k$ for $1 \le k \le 63$. The graphs show the number of iterations required to reduce the norm of the initial error by a factor of 100 for each $\mathbf{w}_k$. Note that for $\omega = \frac{2}{3}$, the damping is strongest for the oscillatory modes ($32 \le k \le 63$).*

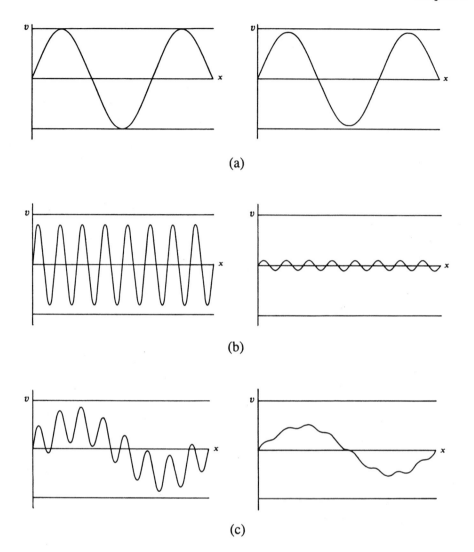

Figure 2.9: *Weighted Jacobi method with* $\omega = \frac{2}{3}$ *applied to the one-dimensional model problem with* $n = 64$ *points and with an initial guess consisting of* (a) $\mathbf{w}_3$, (b) $\mathbf{w}_{16}$, *and* (c) $(\mathbf{w}_2 + \mathbf{w}_{16})/2$. *The figures show the approximation after one iteration (left side) and after* 10 *iterations (right side).*

This distinction is illustrated in Fig. 2.11. As before, the wavenumber k is plotted against the number of iterations required to reduce the norm of the initial error by a factor of 100. In Fig. 2.11(a), the initial guess (and error) consists of the eigenvectors of R_G with wavenumbers $1 \leq k \leq 63$. The graph looks similar to the eigenvalue graph of Fig. 2.10. In Fig. 2.11(b), the initial guess consists of the eigenvectors of the original matrix A. The structure of this graph would be much more difficult to anticipate analytically. We see that when convergence of the Gauss–Seidel method is described in terms of the modes of A, then once again the smooth modes are damped slowly, while the oscillatory modes show rapid decay.

We have looked in detail at the convergence properties of some basic relaxation schemes. The experiments we presented reflect the experience of many practi-

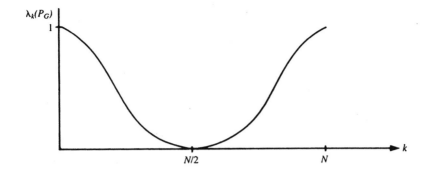

Figure 2.10: *Eigenvalues of the Gauss–Seidel iteration matrix. The eigenvalues* $\lambda_k = \cos^2\left(\frac{k\pi}{n}\right)$ *are plotted as if k were a continuous variable on the interval* $0 \le k \le n$.

tioners. These schemes work very well for the first several iterations. Inevitably, however, convergence slows and the entire scheme appears to stall. We have found a simple explanation for this phenomenon: the rapid decrease in error during the early iterations is due to the efficient elimination of the oscillatory modes of that error; but once the oscillatory modes have been removed, the iteration is much less effective in reducing the remaining smooth components.

There is also a good physical explanation for why smooth error modes are so resistant to relaxation. Recall from (2.2) that stationary linear iterations can be written in the form

$$\mathbf{v}^{(1)} = \mathbf{v}^{(0)} + B\mathbf{r}^{(0)}.$$

Subtracting this equation from the exact solution $\mathbf{u}$, the error at the next step is

$$\mathbf{e}^{(1)} = \mathbf{e}^{(0)} - B\mathbf{r}^{(0)}.$$

We see that changes in the error are made with *spatially local* corrections expressed through the residual. If the residual is small relative to the error itself, then changes in the error will be correspondingly small. At least for the model problems we have posed, smooth error modes have relatively small residuals (Exercise 19), so the error decreases slowly. Conversely, oscillatory errors tend to have relatively large residuals and the corrections to the error with a single relaxation sweep can be significant.

Many relaxation schemes possess this property of eliminating the oscillatory modes and leaving the smooth modes. This so-called *smoothing property* is a serious limitation of conventional relaxation methods. However, this limitation can be overcome and the remedy is one of the pathways to multigrid.

In one very brief chapter, we have barely touched upon the wealth of lore and theory surrounding iterative methods. The subject constitutes a large and important domain of classical numerical analysis. It is also filled with very elegant mathematics from both linear algebra and analysis. However, esoteric iterative methods are not required for the development of multigrid. The most effective multigrid techniques are usually built upon the simple relaxation schemes presented in this chapter. We now use these few basic schemes and develop them into far more powerful methods.

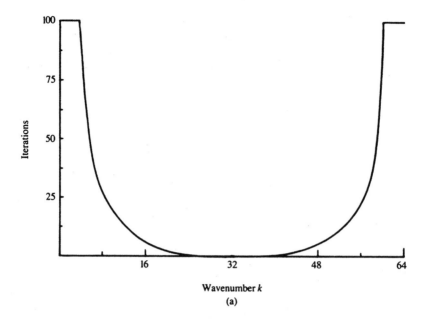

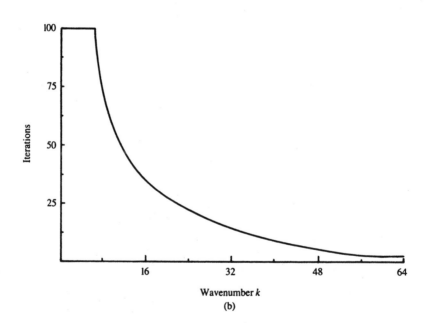

Figure 2.11: *Gauss–Seidel iteration matrix applied to the model problem with $n = 64$ points. The initial guesses consist of* (a) *the eigenvectors of the iteration matrix R_G with wavenumbers $1 \le k \le 63$ and* (b) *the eigenvectors of A with wavenumbers $1 \le k \le 63$. The figure shows the number of iterations required to reduce the norm of the initial error by a factor of 100 for each initial guess.*

Exercises

1. **Residual vs. error.** Consider the two systems of linear equations given in the box on residuals and errors in this chapter. Make a sketch showing the pair of lines represented by each system. Mark the exact solution **u** and the approximation **v**. Explain why, even though the error is the same in both cases, the residual is small in one case and large in the other.

2. **Residual equation.** Use the definition of the algebraic error and the residual to derive the residual equation $A\mathbf{e} = \mathbf{r}$.

3. **Weighted Jacobi iteration.**

   (a) Starting with the component form of the weighted Jacobi method, show that it can be written in matrix form as $\mathbf{v}^{(1)} = [(1-\omega)I + \omega R_J]\mathbf{v}^{(0)} + \omega D^{-1}\mathbf{f}$.

   (b) Show that the weighted Jacobi method may also be written in the form
   $$\mathbf{v}^{(1)} = R_\omega \mathbf{v}^{(0)} + \omega D^{-1}\mathbf{f}.$$

   (c) Show that the weighted Jacobi iteration may also be expressed in the form
   $$\mathbf{v}^{(1)} = \mathbf{v}^{(0)} + \omega D^{-1}\mathbf{r}^{(0)},$$
   where $\mathbf{r}^{(0)}$ is the residual associated with the approximation $\mathbf{v}^{(0)}$.

   (d) Assume that A is the matrix associated with the model problem. Show that the weighted Jacobi iteration matrix can be expressed as
   $$R_\omega = I - \frac{\omega}{2}A.$$

4. **General stationary linear iteration.** It was shown that a general stationary linear iteration can be expressed in the form
   $$\mathbf{v}^{(1)} = (I - BA)\mathbf{v}^{(0)} + B\mathbf{f} \equiv R\mathbf{v}^{(0)} + B\mathbf{f}.$$

   (a) Show that m sweeps of the iteration has the form
   $$\mathbf{v}^{(1)} = R^m \mathbf{v}^{(0)} + C(\mathbf{f}).$$
   Find an expression for $C(\mathbf{f})$.

   (b) Show that the form of the iteration given above is equivalent to
   $$\mathbf{v}^{(1)} = \mathbf{v}^{(0)} + B\mathbf{r}^{(0)},$$
   where $\mathbf{r}^{(0)}$ is the initial residual. Use this form to argue that the exact solution to the linear system, **u**, is unchanged by (and is therefore a fixed point of) the iteration.

5. **Interpreting Gauss–Seidel.** Show that the Gauss–Seidel iteration is equivalent to successively setting each component of the residual to zero.

6. **Zero right side.** Argue that in analyzing the error in a stationary linear relaxation scheme applied to $A\mathbf{u} = \mathbf{f}$, it is sufficient to consider $A\mathbf{u} = \mathbf{0}$ with arbitrary initial guesses.

7. **Asymptotic convergence rate.** Explain why the asymptotic convergence rate,

$$- \log_{10} \rho(R),$$

is positive. Which iteration matrix gives a higher asymptotic convergence rate: one with $\rho(R) = 0.1$ or one with $\rho(R) = 0.9$? Explain.

8. **Eigenvalues of the model problem.** Compute the eigenvalues of the matrix A of the one-dimensional model problem. (Hint: Write out a typical equation of the system $A\mathbf{w} = \lambda\mathbf{w}$ with $w_0 = w_n = 0$. Notice that vectors of the form $w_j = \sin\left(\frac{jk\pi}{n}\right)$, $1 \le k \le n - 1$, $0 \le j \le n$, satisfy the boundary conditions.) How many distinct eigenvalues are there? Compute $\lambda_1, \lambda_2, \lambda_{n-2}, \lambda_{n-1}$ when $n = 32$.

9. **Eigenvectors of the model problem.** Using the results of the previous problem, find the eigenvectors of the one-dimensional model problem matrix A.

10. **Jacobi eigenvalues and eigenvectors.** Find the eigenvalues of the weighted Jacobi iteration matrix when it is applied to the one-dimensional model problem matrix A. Verify that the eigenvectors of R_ω are the same as the eigenvectors of A.

11. **Fourier modes.** Consider the interval $0 \le x \le 1$ with grid points $x_j = \frac{j}{n}$, $0 \le j \le n$. Show that the kth Fourier mode $w_{k,j} = \sin\left(\frac{jk\pi}{n}\right)$ has wavelength $\ell = \frac{2}{k}$. Which mode has wavelength $\ell = 8h$? Which mode has wavelength $\ell = \frac{1}{4}$?

12. **Aliasing.** On a grid with $n - 1$ interior points, show that the mode $w_{k,j} = \sin\left(\frac{jk\pi}{n}\right)$ with $n < k < 2n$ is actually represented as the mode $\mathbf{w}_{k'}$ where $k' = 2n - k$. How is the mode with wavenumber $k = \frac{3n}{2}$ represented on the grid? How is the mode with wavelength $l = \frac{4h}{3}$ represented on the grid? Make sketches for these two examples.

13. **Optimal Jacobi.** Show that when the weighted Jacobi method is used with $\omega = \frac{2}{3}$, the smoothing factor is $\frac{1}{3}$. Show that if ω is chosen to damp the smooth modes effectively, then the oscillatory modes are actually amplified.

14. **Gauss–Seidel eigenvalues and eigenvectors.**

   (a) Show that the eigenvalue problem for the Gauss–Seidel iteration matrix, $R_G\mathbf{w} = \lambda\mathbf{w}$, may be expressed in the form $U\mathbf{w} = (D - L)\lambda I\mathbf{w}$, where U, L, D are defined in the text.

   (b) Write out the equations of this system and note the boundary condition $w_0 = w_n = 0$. Look for solutions of this system of equations of the form $w_j = \mu^j$, where $\mu \in \mathbf{C}$ must be determined. Show that the boundary conditions can be satisfied only if $\lambda = \lambda_k = \cos^2\left(\frac{k\pi}{n}\right)$, $1 \le k \le n - 1$.

   (c) Show that the eigenvector associated with λ_k is $w_{k,j} = \cos(\frac{k\pi}{n}) \sin\left(\frac{jk\pi}{n}\right)$.

15. **Richardson iteration.**

   (a) Recall that for real vectors $\mathbf{u}$, $\mathbf{v}$, the inner product is given by $(\mathbf{u}, \mathbf{v}) = \mathbf{u}^T\mathbf{v}$ and $\|\mathbf{u}\|_2^2 = (\mathbf{u}, \mathbf{u})$. Furthermore, if A is symmetric positive definite, then $\|A\|_2 = \rho(A)$, the spectral radius of A. Richardson's iteration is given by

$$\mathbf{v}^{(1)} = \mathbf{v}^{(0)} + \frac{s}{\|A\|_2}\mathbf{r}^{(0)} \quad \text{for } 0 < s < 2,$$

where $\mathbf{r}^{(0)} = \mathbf{f} - A\mathbf{v}^{(0)}$ is the residual. Show that when A has a constant diagonal, this method reduces to the weighted Jacobi method.

(b) Show that the error after one sweep of Richardson's method is governed by

$$\|\mathbf{e}^{(1)}\|_2^2 \leq \left[1 - \frac{s(2 - s)(A\mathbf{e}^{(0)}, \mathbf{e}^{(0)})}{\|A\|_2 \ (\mathbf{e}^{(0)}, \mathbf{e}^{(0)})}\right] \|\mathbf{e}^{(0)}\|_2^2.$$

(c) If the eigenvalues of A are ordered $0 < \lambda_1 < \lambda_2 < \cdots < \lambda_n$ and the smallest eigenvalues correspond to the smooth modes, show that Richardson's method has the smoothing property. (Use the fact that the eigenvalues are given by the Rayleigh quotients of the eigenvectors, $\lambda_k = (A\mathbf{w}_k, \mathbf{w}_k)/(\mathbf{w}_k, \mathbf{w}_k)$, where $\mathbf{w}_k$ is the eigenvector associated with λ_k.)

16. **Properties of Gauss–Seidel.** Assume A is symmetric, positive definite.

(a) Show that the jth step of a *single* sweep of the Gauss–Seidel method applied to $A\mathbf{u} = \mathbf{f}$ may be expressed as

$$v_j \leftarrow v_j + \frac{r_j}{a_{jj}}.$$

(b) Show that the jth step of a *single* sweep of the Gauss–Seidel method can be expressed in vector form as

$$\mathbf{v} \leftarrow \mathbf{v} + \frac{(\mathbf{r}, \hat{\mathbf{e}}_j)}{(A\hat{\mathbf{e}}_j, \hat{\mathbf{e}}_j)} \hat{\mathbf{e}}_j, \quad 1 \leq j \leq n,$$

where $\hat{\mathbf{e}}_j$ is the jth unit vector.

(c) Show that each sweep of Gauss–Seidel decreases the quantity $(A\mathbf{e}, \mathbf{e})$, where $\mathbf{e} = \mathbf{u} - \mathbf{v}$.

(d) Show that Gauss–Seidel is optimal in the sense that the quantity $\|\mathbf{e} - s\hat{\mathbf{e}}_j\|_A$ is minimized for each $1 \leq j \leq n$ when $s = (\mathbf{r}, \hat{\mathbf{e}}_j)/(A\hat{\mathbf{e}}_j, \hat{\mathbf{e}}_j)$, which is precisely a Gauss–Seidel step.

17. **Matrix 2-norm.** Show that the matrix 2-norm is given by

$$\|A\|_2 = \sqrt{\rho(A^T A)}.$$

Use the definition of matrix norm and the relations $\|\mathbf{x}\|_2^2 = (\mathbf{x}, \mathbf{x})$ and $\|A\mathbf{x}\|_2^2 = (A\mathbf{x}, A\mathbf{x})$.

18. **Error and residual norms.** The *condition number* of a matrix, $\text{cond}(A) = \|A\|_2\|A^{-1}\|_2$, gives an idea of how well the residual measures the error. In the following exercise, use the property of matrix and vector norms that $\|Ax\| \leq \|A\|\|x\|$.

(a) Begin with the relations $Ae = r$ and $A^{-1}f = u$. Taking norms and combining terms, show that

$$\frac{\|r\|_2}{\|f\|_2} \leq \text{cond}(A)\frac{\|e\|_2}{\|u\|_2}.$$

Knowing that this bound is sharp (that is, equality can be achieved), interpret this inequality in the case that the condition number is large.

(b) Now begin with the relations $Au = f$ and $A^{-1}r = e$. Taking norms and combining terms, show that

$$\frac{\|e\|_2}{\|u\|_2} \leq \text{cond}(A)\frac{\|r\|_2}{\|f\|_2}.$$

Knowing that this bound is sharp (that is, equality can be achieved), interpret this inequality in the case that the condition number is large.

(c) Combine the above bounds to form the following relations:

$$\frac{1}{\text{cond}(A)}\frac{\|r\|_2}{\|f\|_2} \leq \frac{\|e\|_2}{\|u\|_2} \leq \text{cond}(A)\frac{\|r\|_2}{\|f\|_2}.$$

19. **Residuals of smooth errors.** Consider the residual equation, $A\mathbf{e} = \mathbf{r}$, at a single point, where A is the matrix for the model problem in either one or two dimensions. Show that if $\mathbf{e}$ is smooth (for example, nearly constant), then $\mathbf{r}$ is small relative to $\|A\|\|\mathbf{e}\|$. Conversely, show that if $\mathbf{e}$ is oscillatory, then $\mathbf{r}$ is relatively large.

20. **Numerical experiments.** Write a short program that performs weighted Jacobi (with variable ω), Gauss–Seidel, and red-black Gauss–Seidel for the one-dimensional model problem. First reproduce the experiments shown in Fig. 2.3. Then experiment with initial guesses with different wavenumbers. Describe how each method performs as the wavenumbers increase and approach n.

Chapter 3

Elements of Multigrid

Through analysis and experimentation, we have examined some of the basic iterative methods. Our discoveries have formed the beginnings of what we might call a spectral (or Fourier mode) picture of relaxation schemes. As we proceed, more essential details of this picture will become clear. So far we have established that many standard iterative methods possess the smoothing property. This property makes these methods very effective at eliminating the high-frequency or oscillatory components of the error, while leaving the low-frequency or smooth components relatively unchanged. The immediate issue is whether these methods can be modified in some way to make them effective on all error components.

One way to improve a relaxation scheme, at least in its early stages, is to use a good initial guess. A well-known technique for obtaining an improved initial guess is to perform some preliminary iterations on a coarse grid. Relaxation on a coarse grid is less expensive because there are fewer unknowns to be updated. Also, because the convergence factor behaves like $1 - O(h^2)$, the coarse grid will have a marginally improved convergence rate. This line of reasoning at least suggests that coarse grids might be worth considering.

With the coarse grid idea in mind, we can think more carefully about its implications. Recall that most basic relaxation schemes suffer in the presence of smooth components of the error. Assume that a particular relaxation scheme has been applied until only smooth error components remain. We now ask what these smooth components look like on a coarser grid. Figure 3.1 shows the answer. A smooth wave with $k = 4$ on a grid Ω^h with $n = 12$ points has been projected directly to the grid Ω^{2h} with $n = 6$ points. On this coarse grid, the original wave still has a wavenumber of $k = 4$. We see that a smooth wave on Ω^h looks *more oscillatory* on Ω^{2h}.

To be more precise, note that the grid points of the coarse grid Ω^{2h} are the even-numbered grid points of the fine grid Ω^h. Consider the kth mode on the fine grid evaluated at the even-numbered grid points. If $1 \le k < \frac{n}{2}$, its components may be written as

$$w^h_{k,2j} = \sin\left(\frac{2jk\pi}{n}\right) = \sin\left(\frac{jk\pi}{n/2}\right) = w^{2h}_{k,j}, \qquad 1 \le k < \frac{n}{2}.$$

Notice that superscripts have been used to indicate the grids on which the vectors are defined. From this identity, we see that the kth mode on Ω^h becomes the kth

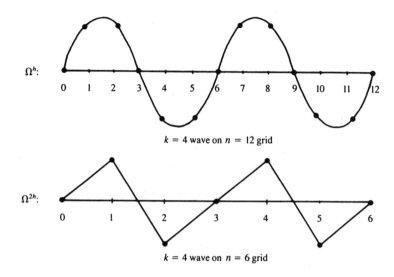

Figure 3.1: *Wave with wavenumber $k = 4$ on Ω^h ($n = 12$ points) projected onto Ω^{2h} ($n = 6$ points). The coarse grid "sees" a wave that is more oscillatory on the coarse grid than on the fine grid.*

mode on Ω^{2h}; this fact is easier to understand by noting that there are half as many modes on Ω^{2h} as there are on Ω^h. The important consequence of this fact is that in passing from the fine grid to the coarse grid, a mode becomes more oscillatory. This is true provided that $1 \leq k < \frac{n}{2}$. It should be verified that the $k = \frac{n}{2}$ mode on Ω^h becomes the zero vector on Ω^{2h}.

As an aside, it is worth mentioning that fine-grid modes with $k > \frac{n}{2}$ undergo a more curious transformation. Through the phenomenon of *aliasing* mentioned earlier, the kth mode on Ω^h becomes the $(n - k)$th mode on Ω^{2h} when $k > \frac{n}{2}$ (Exercise 1). In other words, the oscillatory modes of Ω^h are misrepresented as relatively smooth modes on Ω^{2h}.

The important point is that smooth modes on a fine grid look less smooth on a coarse grid. This suggests that when relaxation begins to stall, signaling the predominance of smooth error modes, it is advisable to move to a coarser grid; there, the smooth error modes appear more oscillatory and relaxation will be more effective. The question is: how do we move to a coarser grid and relax on the more oscillatory error modes?

It is at this point that multigrid begins to come together like a jigsaw puzzle. We must keep all of the related facts in mind. Recall that we have an equation for the error itself, namely, the residual equation. If $\mathbf{v}$ is an approximation to the exact solution $\mathbf{u}$, then the error $\mathbf{e} = \mathbf{u} - \mathbf{v}$ satisfies

$$A\mathbf{e} = \mathbf{r} = \mathbf{f} - A\mathbf{v},$$

which says that we can relax directly on the error by using the residual equation. There is another argument that justifies the use of the residual equation:

Relaxation on the original equation $A\mathbf{u} = \mathbf{f}$ with an arbitrary initial guess $\mathbf{v}$ is equivalent to relaxing on the residual equation $A\mathbf{e} = \mathbf{r}$ with the specific initial guess $\mathbf{e} = \mathbf{0}$.

This intimate connection between the original and the residual equations further motivates the use of the residual equation (Exercise 2).

We must now gather these loosely connected ideas. We know that many relaxation schemes possess the smoothing property. This leads us to consider using coarser grids during the computation to focus the relaxation on the oscillatory components of the error. In addition, there seems to be good reason to involve the residual equation in the picture. We now try to give these ideas a little more definition by proposing two strategies.

We begin by proposing a strategy that uses coarse grids to obtain better initial guesses.

- Relax on $A\mathbf{u} = \mathbf{f}$ on a very coarse grid to obtain an initial guess for the next finer grid.

  ⋮

- Relax on $A\mathbf{u} = \mathbf{f}$ on Ω^{4h} to obtain an initial guess for Ω^{2h}.

- Relax on $A\mathbf{u} = \mathbf{f}$ on Ω^{2h} to obtain an initial guess for Ω^{h}.

- Relax on $A\mathbf{u} = \mathbf{f}$ on Ω^{h} to obtain a final approximation to the solution.

This idea of using coarser grids to generate improved initial guesses is the basis of a strategy called *nested iteration*. Although the approach is attractive, it also leaves some questions. For instance, what does it mean to relax on $A\mathbf{u} = \mathbf{f}$ on Ω^{2h}? We must somehow define the original problem on the coarser grids. Also, what happens if, having once reached the fine grid, there are still smooth components in the error? We may have obtained some improvement by using the coarse grids, but the final iteration will stall if smooth components still remain. We return to these questions and find answers that will allow us to use nested iteration in a very powerful way.

A second strategy incorporates the idea of using the residual equation to relax on the error. It can be represented by the following procedure:

- Relax on $A\mathbf{u} = \mathbf{f}$ on Ω^{h} to obtain an approximation $\mathbf{v}^{h}$.

- Compute the residual $\mathbf{r} = \mathbf{f} - A\mathbf{v}^{h}$.
  Relax on the residual equation $A\mathbf{e} = \mathbf{r}$ on Ω^{2h} to obtain
  an approximation to the error $\mathbf{e}^{2h}$.

- Correct the approximation obtained on Ω^{h} with the error estimate obtained on Ω^{2h} : $\mathbf{v}^{h} \leftarrow \mathbf{v}^{h} + \mathbf{e}^{2h}$.

This procedure is the basis of what is called *the correction scheme*. Having relaxed on the fine grid until convergence deteriorates, we relax on the *residual* equation on a coarser grid to obtain an approximation to the error itself. We then return to the fine grid to correct the approximation first obtained there.

There is a rationale for using this correction strategy, but it also leaves some questions to be answered. For instance, what does it mean to relax on $A\mathbf{e} = \mathbf{r}$ on Ω^{2h}? To answer this question, we first need to know how to compute the residual

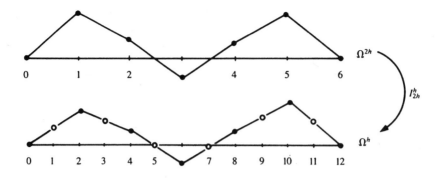

Figure 3.2: *Interpolation of a vector on coarse grid Ω^{2h} to fine grid Ω^h.*

on Ω^h and transfer it to Ω^{2h}. We also need to know how to relax on Ω^{2h} and what initial guess should be used. Moreover, how do we transfer the error estimate from Ω^{2h} back to Ω^h? These questions suggest that we need mechanisms for transferring information between the grids. We now turn to this important consideration.

In our discussion of intergrid transfers, we consider only the case in which the coarse grid has twice the grid spacing of the next finest grid. This is a nearly universal practice, because there is usually no advantage in using grid spacings with ratios other than 2. Think for a moment about the step in the correction scheme that requires transferring the error approximation $\mathbf{e}^{2h}$ from the coarse grid Ω^{2h} to the fine grid Ω^h. This is a common procedure in numerical analysis and is generally called *interpolation* or *prolongation*. Many interpolation methods could be used. Fortunately, for most multigrid purposes, the simplest of these is quite effective. For this reason, we consider only linear interpolation.

The linear interpolation operator will be denoted I_{2h}^h. It takes coarse-grid vectors and produces fine-grid vectors according to the rule $I_{2h}^h \mathbf{v}^{2h} = \mathbf{v}^h$, where

$$v_{2j}^h = v_j^{2h},$$
$$v_{2j+1}^h = \frac{1}{2}\left(v_j^{2h} + v_{j+1}^{2h}\right), \quad 0 \le j \le \frac{n}{2} - 1.$$

Figure 3.2 shows graphically the action of I_{2h}^h. At even-numbered fine-grid points, the values of the vector are transferred directly from Ω^{2h} to Ω^h. At odd-numbered fine-grid points, the value of $\mathbf{v}^h$ is the average of the adjacent coarse-grid values.

In anticipation of discussions to come, we note that I_{2h}^h is a linear operator from $\mathbf{R}^{\frac{n}{2}-1}$ to $\mathbf{R}^{n-1}$. It has full rank and the trivial null space, $\mathcal{N} = \{0\}$. For the case $n = 8$, this operator has the form

$$I_{2h}^h \mathbf{v}^{2h} = \frac{1}{2}\begin{bmatrix} 1 & & \\ 2 & & \\ 1 & 1 & \\ & 2 & \\ & 1 & 1 \\ & & 2 \\ & & 1 \end{bmatrix}\begin{bmatrix} v_1 \\ v_2 \\ v_3 \end{bmatrix}_{2h} = \begin{bmatrix} v_1 \\ v_2 \\ v_3 \\ v_4 \\ v_5 \\ v_6 \\ v_7 \end{bmatrix}_h = \mathbf{v}^h.$$

How well does this interpolation process work? First assume that the "real" error (which is not known exactly) *is* a smooth vector on the fine grid. Assume

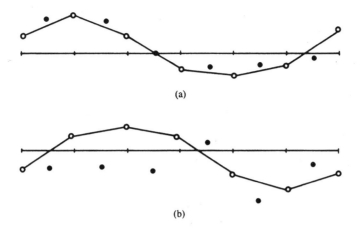

Figure 3.3: (a) *If the exact error on Ω^h (indicated by $\circ$ and $\bullet$) is smooth, an interpolant of the coarse-grid error $\mathbf{e}^{2h}$ (solid line connecting $\circ$ points) should give a good representation of the exact error.* (b) *If the exact error on Ω^h (indicated by $\circ$ and $\bullet$) is oscillatory, an interpolant of the coarse-grid error $\mathbf{e}^{2h}$ (solid line connecting $\circ$ points) may give a poor representation of the exact error.*

also that a coarse-grid approximation to the error has been determined on Ω^{2h} and that this approximation is exact at the coarse-grid points. When this coarse-grid approximation is interpolated to the fine grid, the interpolant is also smooth. Therefore, we expect a relatively good approximation to the fine-grid error, as shown in Fig. 3.3(a). By contrast, if the "real" error is oscillatory, even a very good coarse-grid approximation may produce an interpolant that is not very accurate. This situation is shown in Fig. 3.3(b).

Thus, interpolation is most effective when the error is smooth. Because interpolation is necessary for both nested iteration and the correction scheme, we may conclude that these two processes are most effective when the error is smooth. As we will see shortly, these processes provide a fortunate complement to relaxation, which is most effective when the error is oscillatory.

For two-dimensional problems, the interpolation operator may be defined in a similar way. If we let $I_{2h}^h \mathbf{v}^{2h} = \mathbf{v}^h$, then the components of $\mathbf{v}^h$ are given by

$$
\begin{aligned}
v_{2i,2j}^h &= v_{ij}^{2h}, \\
v_{2i+1,2j}^h &= \frac{1}{2}\left(v_{ij}^{2h} + v_{i+1,j}^{2h}\right), \\
v_{2i,2j+1}^h &= \frac{1}{2}\left(v_{ij}^{2h} + v_{i,j+1}^{2h}\right), \\
v_{2i+1,2j+1}^h &= \frac{1}{4}\left(v_{ij}^{2h} + v_{i+1,j}^{2h} + v_{i,j+1}^{2h} + v_{i+1,j+1}^{2h}\right), \quad 0 \le i,j \le \frac{n}{2} - 1.
\end{aligned}
$$

The second class of intergrid transfer operations involves moving vectors from a fine grid to a coarse grid. They are generally called *restriction* operators and are denoted by I_h^{2h}. The most obvious restriction operator is *injection*. It is defined by $I_h^{2h} \mathbf{v}^h = \mathbf{v}^{2h}$, where

$$
v_j^{2h} = v_{2j}^h.
$$

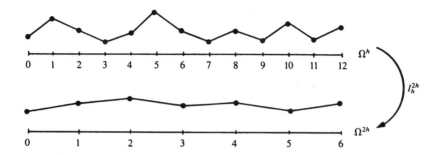

Figure 3.4: *Restriction by full weighting of a fine-grid vector to the coarse grid.*

In other words, with injection, the coarse-grid vector simply takes its value directly from the corresponding fine-grid point.

An alternate restriction operator, called *full weighting*, is defined by $I_h^{2h}\mathbf{v}^h = \mathbf{v}^{2h}$, where

$$v_j^{2h} = \frac{1}{4}\left(v_{2j-1}^h + 2v_{2j}^h + v_{2j+1}^h\right), \qquad 1 \le j \le \frac{n}{2} - 1.$$

As Fig. 3.4 shows, the values of the coarse-grid vector are weighted averages of values at neighboring fine-grid points.

In the discussion that follows, we use full weighting as a restriction operator. However, in some instances, injection may be the better choice. The issue of inter-grid transfers, which is an important part of multigrid theory, is discussed at some length in Brandt's guide to multigrid [4].

The full weighting operator is a linear operator from $\mathbf{R}^{n-1}$ to $\mathbf{R}^{\frac{n}{2}-1}$. It has a rank of $\frac{n}{2} - 1$ (Exercise 4) and a null space of dimension $\frac{n}{2}$ (Exercise 5). For the case $n = 8$, the full weighting operator has the form

$$I_h^{2h}\mathbf{v}^h = \frac{1}{4}\begin{bmatrix} 1 & 2 & 1 & & & & \\ & & 1 & 2 & 1 & & \\ & & & & 1 & 2 & 1 \end{bmatrix}\begin{bmatrix} v_1 \\ v_2 \\ v_3 \\ v_4 \\ v_5 \\ v_6 \\ v_7 \end{bmatrix}_h = \begin{bmatrix} v_1 \\ v_2 \\ v_3 \end{bmatrix}_{2h} = \mathbf{v}^{2h}.$$

One reason for our choice of full weighting as a restriction operator is the important fact (Exercise 6) that

$$I_{2h}^h = c(I_h^{2h})^T, \qquad c \in \mathbf{R}.$$

The fact that the interpolation operator and the full weighting operator are trans-poses of each other up to a constant is called a *variational property* and will soon be of importance.

For the sake of completeness, we give the full weighting operator in two dimensions. It is just an averaging of the fine-grid nearest neighbors. Letting $I_h^{2h}\mathbf{v}^h = \mathbf{v}^{2h}$, we have that

$$\begin{aligned} v_{ij}^{2h} = {} & \frac{1}{16}\big[v_{2i-1,2j-1}^h + v_{2i-1,2j+1}^h + v_{2i+1,2j-1}^h + v_{2i+1,2j+1}^h \\ & + 2\big(v_{2i,2j-1}^h + v_{2i,2j+1}^h + v_{2i-1,2j}^h + v_{2i+1,2j}^h\big) \\ & + 4v_{2i,2j}^h\big], \quad 1 \le i,j \le \frac{n}{2} - 1. \end{aligned}$$

We now have a well-defined way to transfer vectors between fine and coarse grids. Therefore, we can return to the correction scheme and make it precise. To do this, we define the following two-grid correction scheme.

Two-Grid Correction Scheme

$$\mathbf{v}^h \leftarrow MG(\mathbf{v}^h, \mathbf{f}^h).$$

- Relax ν_1 times on $A^h \mathbf{u}^h = \mathbf{f}^h$ on Ω^h with initial guess $\mathbf{v}^h$.

- Compute the fine-grid residual $\mathbf{r}^h = \mathbf{f}^h - A^h \mathbf{v}^h$ and restrict it to the coarse grid by $\mathbf{r}^{2h} = I_h^{2h} r^h$.

- Solve $A^{2h} \mathbf{e}^{2h} = \mathbf{r}^{2h}$ on Ω^{2h}.

- Interpolate the coarse-grid error to the fine grid by $\mathbf{e}^h = I_{2h}^h \mathbf{e}^{2h}$ and correct the fine-grid approximation by $\mathbf{v}^h \leftarrow \mathbf{v}^h + \mathbf{e}^h$.

- Relax ν_2 times on $A^h \mathbf{u}^h = \mathbf{f}^h$ on Ω^h with initial guess $\mathbf{v}^h$.

This procedure is simply the original correction scheme, now refined by the use of the intergrid transfer operators. We relax on the fine grid until it ceases to be worthwhile; in practice, ν_1 is often 1, 2, or 3. The residual of the current approximation is computed on Ω^h and then transferred by a restriction operator to the coarse grid. As it stands, the procedure calls for the exact solution of the residual equation on Ω^{2h}, which may not be possible. However, if the coarse-grid error can at least be approximated, it is then interpolated up to the fine grid, where it is used to correct the fine-grid approximation. This is followed by ν_2 additional fine-grid relaxation sweeps.

Several comments are in order. First, notice that the superscripts h or $2h$ are essential to indicate the grid on which a particular vector or matrix is defined. Second, all of the quantities in the above procedure are well defined except for A^{2h}. For the moment, we take A^{2h} simply to be the result of discretizing the problem on Ω^{2h}. Finally, the integers ν_1 and ν_2 are parameters in the scheme that control the number of relaxation sweeps before and after visiting the coarse grid. They are usually fixed at the start, based on either theoretical considerations or on past experimental results.

It is important to appreciate the complementarity at work in the process. Relaxation on the fine grid eliminates the oscillatory components of the error, leaving a relatively smooth error. Assuming the residual equation can be solved accurately on Ω^{2h}, it is still important to transfer the error accurately back to the fine grid. Because the error is smooth, interpolation should work very well and the correction of the fine-grid solution should be effective.

Numerical example. A numerical example will be helpful. Consider the weighted Jacobi method with $\omega = \frac{2}{3}$ applied to the one-dimensional model problem $A\mathbf{u} = \mathbf{0}$ on a grid with $n = 64$ points. We use an initial guess,

$$v_j^h = \frac{1}{2}\left[\sin\left(\frac{16j\pi}{n}\right) + \sin\left(\frac{40j\pi}{n}\right)\right],$$

consisting of the $k = 16$ and $k = 40$ modes. The following two-grid correction scheme is used:

- Relax three times on $A^h \mathbf{u}^h = \mathbf{0}$ on Ω^h with initial guess $\mathbf{v}^h$.

- Compute $\mathbf{r}^{2h} = I_h^{2h} \mathbf{r}^h$.

- Relax three times on $A^{2h} \mathbf{e}^{2h} = \mathbf{r}^{2h}$ on Ω^{2h} with initial guess $\mathbf{e}^{2h} = \mathbf{0}$.

- Correct the fine-grid approximation: $\mathbf{v}^h \leftarrow \mathbf{v}^h + I_{2h}^h \mathbf{e}^{2h}$.

- Relax three times on $A^h \mathbf{u}^h = \mathbf{0}$ on Ω^h with initial guess $\mathbf{v}^h$.

- Compute $\mathbf{r}^{2h} = I_h^{2h} \mathbf{r}^h$.

- Relax three times on $A^{2h} \mathbf{e}^{2h} = \mathbf{r}^{2h}$ on Ω^{2h} with initial guess $\mathbf{e}^{2h} = \mathbf{0}$.

- Correct the fine-grid approximation: $\mathbf{v}^h \leftarrow \mathbf{v}^h + I_{2h}^h \mathbf{e}^{2h}$.

The results of this calculation are given in Fig. 3.5. The initial guess with its two modes is shown in the top left figure. In the top right, the approximation $\mathbf{v}^h$ after one relaxation sweep is superimposed on the initial guess. Much of the oscillatory component of the initial guess has already been removed, and the 2-norm of the error has been diminished to 57% of the norm of the initial error. The middle left plot shows the approximation after three relaxation sweeps on the fine grid, again superimposed on the initial guess. The solution (in this case, the error) has become smoother and its norm is now 36% of the initial error norm. Further relaxations on the fine grid would provide only a slow improvement at this point. This signals that it is time to move to the coarse grid.

The middle right plot shows the fine-grid error after one relaxation sweep on the coarse-grid residual equation, superimposed on the initial guess. Clearly, we have achieved another reduction in the error by moving to the coarse grid; the norm of the error is now 26% of the initial error norm. This improvement occurs because the smooth error components, inherited from the fine grid, appear oscillatory on the coarse grid and are quickly removed. The error after three coarse-grid relaxation sweeps is shown in the bottom left figure. The norm of the error is now about 8% of its initial value.

The coarse-grid approximation to the error is now used to correct the fine-grid approximation. After three additional fine-grid relaxations, the 2-norm of the error is reduced to about 3% of the initial error norm. This result is plotted in the bottom right figure. The residual is once again transferred to the coarse grid and three coarse-grid relaxations follow. At this point, the 2-norm of the error is about 1% of its original value. This experiment demonstrates that relaxation, when done on two grids and applied to both the original and the residual equation, can be very powerful.   ⋄⋄

The two-grid correction scheme, as outlined above, leaves one looming procedural question: what is the best way to solve the coarse-grid problem $A^{2h} \mathbf{e}^{2h} = \mathbf{r}^{2h}$? The answer may be apparent, particularly to those who think recursively. The coarse-grid problem is not much different from the original problem. Therefore, we can apply the two-grid correction scheme to the residual equation on Ω^{2h}, which means relaxing there and then moving to Ω^{4h} for the correction step. We can repeat this process on successively coarser grids until a direct solution of the residual equation is possible.

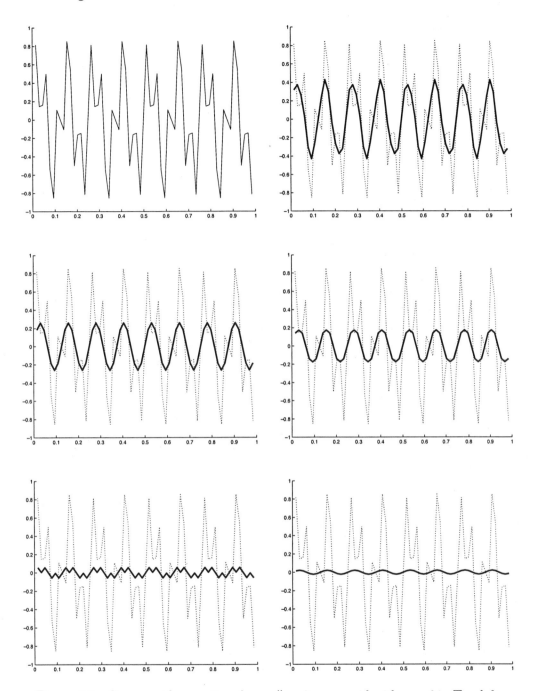

Figure 3.5: *Coarse-grid correction for $-u'' = 0$ on a grid with $n = 64$. Top left: The initial guess, $(\mathbf{w}_{16} + \mathbf{w}_{40})/2$. Top right: The error after one sweep of weighted Jacobi. Middle left: The error after three sweeps of weighted Jacobi. Middle right: The fine-grid error after one sweep of weighted Jacobi on the coarse-grid problem. Bottom left: The fine-grid error after three sweeps of weighted Jacobi on the coarse-grid problem. Bottom right: The fine-grid error after the coarse-grid correction is followed by three weighted Jacobi sweeps on the fine grid.*

To facilitate the description of this procedure, some economy of notation is desirable. The same notation is used for the computer implementation of the resulting algorithm. We call the right-side vector of the residual equation $\mathbf{f}^{2h}$, rather than $\mathbf{r}^{2h}$, because it is just another right-side vector. Instead of calling the solution of the residual equation $\mathbf{e}^{2h}$, we use $\mathbf{u}^{2h}$ because it is just a solution vector. We can then use $\mathbf{v}^{2h}$ to denote approximations to $\mathbf{u}^{2h}$. These changes simplify the notation, but it is still important to remember the meaning of these variables.

One more point needs to be addressed: what initial guess do we use for $\mathbf{v}^{2h}$ on the first visit to Ω^{2h}? Because there is presumably no information available about the solution, $\mathbf{u}^{2h}$, we simply choose $\mathbf{v}^{2h} = 0$. Here then is the two-grid correction scheme, now imbedded within itself. We assume that there are $l > 1$ grids with grid spacings $h, 2h, 4h, \ldots, Lh = 2^{l-1}h$.

V-Cycle Scheme

$$\mathbf{v}^h \leftarrow V^h(\mathbf{v}^h, \mathbf{f}^h)$$

- Relax on $A^h\mathbf{u}^h = \mathbf{f}^h$ ν_1 times with initial guess $\mathbf{v}^h$.
- Compute $\mathbf{f}^{2h} = I_h^{2h}\mathbf{r}^h$.
  - Relax on $A^{2h}\mathbf{u}^{2h} = \mathbf{f}^{2h}$ ν_1 times with initial guess $\mathbf{v}^{2h} = 0$.
  - Compute $\mathbf{f}^{4h} = I_{2h}^{4h}\mathbf{r}^{2h}$.
    - Relax on $A^{4h}\mathbf{u}^{4h} = \mathbf{f}^{4h}$ ν_1 times with initial guess $\mathbf{v}^{4h} = 0$.
    - Compute $\mathbf{f}^{8h} = I_{4h}^{8h}\mathbf{r}^{4h}$.

$$\vdots$$

- Solve $A^{Lh}\mathbf{u}^{Lh} = \mathbf{f}^{Lh}$.

$$\vdots$$

- Correct $\mathbf{v}^{4h} \leftarrow \mathbf{v}^{4h} + I_{8h}^{4h}\mathbf{v}^{8h}$.
    - Relax on $A^{4h}\mathbf{u}^{4h} = \mathbf{f}^{4h}$ ν_2 times with initial guess $\mathbf{v}^{4h}$.
  - Correct $\mathbf{v}^{2h} \leftarrow \mathbf{v}^{2h} + I_{4h}^{2h}\mathbf{v}^{4h}$.
    - Relax on $A^{2h}\mathbf{u}^{2h} = \mathbf{f}^{2h}$ ν_2 times with initial guess $\mathbf{v}^{2h}$.
- Correct $\mathbf{v}^h \leftarrow \mathbf{v}^h + I_{2h}^h\mathbf{v}^{2h}$.
- Relax on $A^h\mathbf{u}^h = \mathbf{f}^h$ ν_2 times with initial guess $\mathbf{v}^h$.

The algorithm telescopes down to the coarsest grid, which can consist of one or a few interior grid points, then works its way back to the finest grid. Figure 3.6(a) shows the schedule for the grids in the order in which they are visited. Because of the pattern in this diagram, this algorithm is called the *V-cycle*. It is our first true multigrid method.

Not surprisingly, the V-cycle has a compact recursive definition, which is given as follows.

V-Cycle Scheme (Recursive Definition)

$$\mathbf{v}^h \leftarrow V^h(\mathbf{v}^h, \mathbf{f}^h).$$

1. Relax ν_1 times on $A^h\mathbf{u}^h = \mathbf{f}^h$ with a given initial guess $\mathbf{v}^h$.

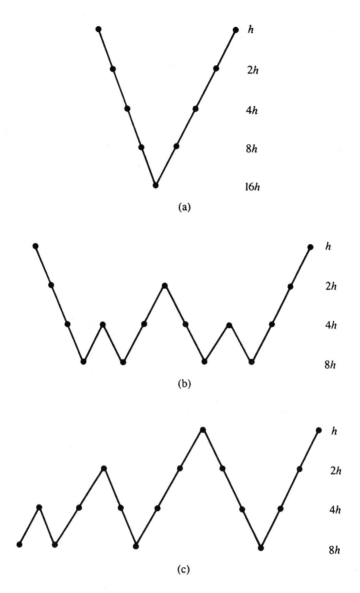

Figure 3.6: *Schedule of grids for* (a) *V-cycle,* (b) *W-cycle, and* (c) *FMG scheme, all on four levels.*

2. If Ω^h = coarsest grid, then go to step 4.

   Else

   $$\mathbf{f}^{2h} \leftarrow I_h^{2h}(\mathbf{f}^h - A^h \mathbf{v}^h),$$
   $$\mathbf{v}^{2h} \leftarrow \mathbf{0},$$
   $$\mathbf{v}^{2h} \leftarrow V^{2h}(\mathbf{v}^{2h}, \mathbf{f}^{2h}).$$

3. Correct $\mathbf{v}^h \leftarrow \mathbf{v}^h + I_{2h}^h \mathbf{v}^{2h}$.

4. Relax ν_2 times on $A^h \mathbf{u}^h = \mathbf{f}^h$ with initial guess $\mathbf{v}^h$.

The V-cycle is just one of a family of multigrid cycling schemes. The entire family is called the μ-cycle method and is defined recursively by the following.

μ-Cycle Scheme

$$\mathbf{v}^h \leftarrow M\mu^h(\mathbf{v}^h, \mathbf{f}^h).$$

1. Relax ν_1 times on $A^h\mathbf{u}^h = \mathbf{f}^h$ with a given initial guess $\mathbf{v}^h$.

2. If Ω^h = coarsest grid, then go to step 4.

   Else
   $$\mathbf{f}^{2h} \leftarrow I_h^{2h}(\mathbf{f}^h - A^h\mathbf{v}^h),$$
   $$\mathbf{v}^{2h} \leftarrow \mathbf{0},$$
   $$\mathbf{v}^{2h} \leftarrow M\mu^{2h}(\mathbf{v}^{2h}, \mathbf{f}^{2h}) \ \mu \text{ times}.$$

3. Correct $\mathbf{v}^h \leftarrow \mathbf{v}^h + I_{2h}^h\mathbf{v}^{2h}$.

4. Relax ν_2 times on $A^h\mathbf{u}^h = \mathbf{f}^h$ with initial guess $\mathbf{v}^h$.

In practice, only $\mu = 1$ (which gives the V-cycle) and $\mu = 2$ are used. Figure 3.6(b) shows the schedule of grids for $\mu = 2$ and the resulting *W-cycle*. We refer to a V-cycle with ν_1 relaxation sweeps before the correction step and ν_2 relaxation sweeps after the correction step as a $V(\nu_1, \nu_2)$-cycle, with a similar notation for W-cycles.

We originally stated that two ideas would lead to multigrid. So far we have developed only the correction scheme. The nested iteration idea has yet to be explored. Recall that nested iteration uses coarse grids to obtain improved initial guesses for fine-grid problems. In looking at the V-cycle, we might ask how to obtain an informed initial guess for the first fine-grid relaxation. Nested iteration would suggest solving a problem on Ω^{2h}. But how can we obtain a good initial guess for the Ω^{2h} problem? Nested iteration sends us to Ω^{4h}. Clearly, we are on another recursive path that leads to the coarsest grid.

The algorithm that joins nested iteration with the V-cycle is called the *full multigrid V-cycle (FMG)*. Given first in explicit terms, it appears as follows.

Full Multigrid V-Cycle

$$\mathbf{v}^h \leftarrow FMG^h(\mathbf{f}^h).$$

Initialize $f^{2h} \leftarrow I_h^{2h}f^h, f^{4h} \leftarrow I_{2h}^{4h}f^{2h}, \ldots.$

- Solve or relax on coarsest grid.

   .

   .

   .

- $\mathbf{v}^{4h} \leftarrow I_{8h}^{4h}\mathbf{v}^{8h}.$
- $\mathbf{v}^{4h} \leftarrow V^{4h}(\mathbf{v}^{4h}, \mathbf{f}^{4h})\,\nu_0$ times.
- $\mathbf{v}^{2h} \leftarrow I_{4h}^{2h}\mathbf{v}^{4h}.$
- $\mathbf{v}^{2h} \leftarrow V^{2h}(\mathbf{v}^{2h}, \mathbf{f}^{2h})\,\nu_0$ times.
- $\mathbf{v}^h \leftarrow I_{2h}^h\mathbf{v}^{2h}.$
- $\mathbf{v}^h \leftarrow V^h(\mathbf{v}^h, \mathbf{f}^h), \nu_0$ times.

We initialize the coarse-grid right sides by transferring $\mathbf{f}^h$ from the fine grid. Another option is to use the original right-side function f. The cycling parameter, ν_0, sets the number of V-cycles done at each level. It is generally determined by a previous numerical experiment; $\nu_0 = 1$ is the most common choice. Expressed recursively, the algorithm has the following compact form.

Full Multigrid V-Cycle (Recursive Form)

$$\mathbf{v}^h \leftarrow FMG^h(\mathbf{f}^h).$$

1. If Ω^h = coarsest grid, set $\mathbf{v}^h \leftarrow \mathbf{0}$ and go to step 3.

    Else

    $$\mathbf{f}^{2h} \leftarrow I_h^{2h}(\mathbf{f}^h),$$
    $$\mathbf{v}^{2h} \leftarrow FMG^{2h}(\mathbf{f}^{2h}).$$

2. Correct $\mathbf{v}^h \leftarrow I_{2h}^h \mathbf{v}^{2h}$.

3. $\mathbf{v}^h \leftarrow V^h(\mathbf{v}^h, \mathbf{f}^h)$ ν_0 times.

Figure 3.6(c) shows the schedule of grids for FMG with $\nu_0 = 1$. Each V-cycle is preceded by a coarse-grid V-cycle designed to provide the best initial guess possible. As we will see, the extra work done in these preliminary V-cycles is not only inexpensive (Exercise 8), but easily pays for itself.

Full multigrid is the complete knot into which the many threads of the preceding chapters are tied. It is a remarkable synthesis of ideas and techniques that individually have been well known and used for a long time. Taken alone, many of these ideas have serious defects. Full multigrid is a technique for integrating them so that they can work together in a way that overcomes these limitations. The result is a very powerful algorithm.

Exercises

1. **Aliasing.** Show that the kth mode on a grid Ω^h with $n - 1$ interior points appears as the $(n - k)$th mode on Ω^{2h} when $\frac{n}{2} < k < n$.

2. **An important equivalence.** Consider a stationary, linear method of the form $\mathbf{v} \leftarrow \mathbf{v} + B^{-1}(\mathbf{f} - A\mathbf{v})$ applied to the problem $A\mathbf{u} = \mathbf{f}$. Use the following steps to show that relaxation on $A\mathbf{u} = \mathbf{f}$ with an arbitrary initial guess is equivalent to relaxation on $A\mathbf{e} = \mathbf{r}$ with the zero initial guess:

    (a) First consider the problem $A\mathbf{u} = \mathbf{f}$ with an arbitrary initial guess $\mathbf{v} = \mathbf{v_0}$. What are the error and residual associated with $\mathbf{v_0}$?

    (b) Now consider the associated residual equation $A\mathbf{e} = \mathbf{r_0} = \mathbf{f} - A\mathbf{v_0}$. What are the error and residual in the initial guess $\mathbf{e_0} = \mathbf{0}$?

    (c) Conclude that the problems in (a) and (b) are equivalent.

3. **Properties of interpolation.** Show that I_{2h}^h based upon linear interpolation is a linear operator with full rank in one and two dimensions.

4. **Properties of restriction.** What is the rank of I_h^{2h} based on (a) full weighting and (b) injection in one and two dimensions?

5. **Null space of full weighting.** Show that the null space of the full weighting operator, $N(I_h^{2h})$, has a basis consisting of vectors of the form

$$(0, 0, \ldots, -1, 2, -1, \ldots, 0, 0)^T.$$

   By counting these vectors, show that the dimension of $N(I_h^{2h})$ is $\frac{n}{2}$.

6. **Variational property.**

   (a) Let I_{2h}^h and I_h^{2h} be defined as in the text. Show that linear interpolation and full weighting satisfy the variational property $I_{2h}^h = c(I_h^{2h})^T$ by computing $c \in \mathbf{R}$ for both one and two dimensions.

   (b) The choice of $c \neq 1$ found in part (a) is used because full weighting essentially preserves constants. Show that, except at the boundary, $I_h^{2h}(\mathbf{1}^h) = \mathbf{1}^{2h}$ (where $\mathbf{1}^h$ and $\mathbf{1}^{2h}$ are the vectors with entries 1 on their respective grids).

7. **Properties of red-black Gauss–Seidel.** Suppose red-black Gauss–Seidel is used with the V-cycle scheme for the one-dimensional model problem.

   (a) Does it matter whether the odd unknowns or even unknowns are updated first? Explain.

   (b) Show that one sweep of red-black Gauss–Seidel on Ω^h leaves the error $\mathbf{e}^h$ in the range of interpolation I_{2h}^h.

   (c) Demonstrate that one V-cycle based on red-black Gauss–Seidel and full weighting is a direct (exact) solver for the one-dimensional model problem.

8. **FMG cost.** The difference in cost between FMG and a single V-cycle is the cost of all but the last V-cycle on Ω^h in the FMG scheme. Estimate the cost of these extra V-cycles. Assume that the cost of a V-cycle on grid Ω^{ph} is proportional to the number of points in that grid, where $p = 2, 4, 8, \ldots, n/2$. Assume also that $\nu_0 = 1$.

Chapter 4

Implementation

The preceding chapter was devoted to the development of several multigrid schemes. We now turn to the practical issues of writing multigrid programs and determining whether they work. This will lead us to issues such as data structures, complexity, predictive tools, diagnostic tools, and performance.

Complexity

Writing multigrid programs can be both fun and challenging. The experience of many practitioners suggests that such programs should be highly modular. This allows them to evolve from simple relaxation programs and makes them much easier to check and debug. Also, the various components of the program (for example, relaxation, interpolation, and restriction subroutines) can be replaced individually.

Choosing a manageable data structure for a multigrid program is essential. Modern programming languages are replete with devices that make data management easy. For example, in most languages, one can declare a *structure* that groups together all the associated information for each grid level. In a structured language, for instance, a V-cycle could be written along the lines of the following *pseudocode*:

```
declare structure:
     grid = { double Ddim_array  f  %% the right hand side
              double Ddim_array  v  %% the current approximation }

declare Grid: array of structure grid

for j = 0 to coarsest - 1
    Grid[j].v <- relax(Grid[j].v, Grid[j].f, num_sweeps_down)
    Grid[j+1].f  <- restrict(Grid[j].f - apply_operator(Grid[j].v))
endfor

Grid[coarsest].v = direct_solve(Grid[coarsest].v, Grid[coarsest].f)

for j = coarsest-1 to 1
    Grid[j].v <- Grid[j].v + interpolate(Grid[j+1].v)
    Grid[j].v <- relax(Grid[j].v, Grid[j].f, num_sweeps_down)
endfor
```

The routines *relax, restrict, apply_operator, interpolate*, and *direct_solve* take the appropriate *Ddim_arrays*, *v* and *f*, for the specified grid level and perform the appropriate operations. We do not describe this type of data management in any further detail, as the advances in these languages occur so rapidly that any discussion would soon be outdated!

We describe a data structure for a simpler FORTRAN-like language. Multigrid codes "grew up" in such an environment and many people learn to write multigrid codes using *MATLAB* or a similar prototyping language with more restrictive data structures. With these languages, there seems to be general agreement that the solutions and right-side vectors on the various grids should be stored contiguously in single arrays. Complicating factors such as irregular domains or local fine-grid patches might require an exception to this practice. However, single arrays are advisable for the regular grids discussed in this chapter.

We begin by considering a four-level V-cycle applied to a one-dimensional problem with $n = 16$ points. A typical data structure is shown in Fig. 4.1. It is instructive to note how the data structure changes as the V-cycle progresses. Each grid needs two arrays: one to hold the current approximations on each grid and one to hold the right-side vectors on each grid. Because boundary values must also be stored, the coarsest grid involves three grid points (one interior and two boundary points). In general, the ℓth coarsest grid involves $2^\ell + 1$ points.

Initially, the entire solution array **v** may be set to zero, which will be the initial guess on each grid. The right-side array **f** will also be set to zero, except for the values on the finest grid, which are known at the outset.

As the V-cycle "descends" into coarser grids, relaxation fills the segment of the solution array corresponding to each grid. At the same time, the residual vectors **f** fill the right-side array corresponding to the next coarsest grid. As the V-cycle "ascends" through finer grids, the right-side array does not change. However, the solution array is overwritten by additional relaxations on each level. Notice that when a new approximation is computed on one level, the approximation on the previous level is zeroed out. This provides a zero initial guess on each level in case another V-cycle is performed.

We now turn to the important questions of complexity. How much do the multigrid schemes cost in terms of storage and computation? The storage question is easier to answer. Consider a d-dimensional grid with n^d points. (Actually, for Dirichlet boundary conditions, there will be $(n-1)^d$ interior points and, as we will see later, for Neumann boundary conditions, there will be $(n+1)^d$ unknown points.) For simplicity, suppose n is a power of 2. We have just seen that two arrays must be stored on each level. The finest grid, Ω^h, requires $2n^d$ storage locations; Ω^{2h} requires 2^{-d} times as much storage as Ω^h; Ω^{4h} requires 4^{-d} times as much storage as Ω^h; in general, Ω^{ph} requires p^{-d} times as much storage as Ω^h. Adding these terms and using the sum of the geometric series as an upper bound gives

$$\text{Storage} = 2n^d\{1 + 2^{-d} + 2^{-2d} + \cdots + 2^{-nd}\} < \frac{2n^d}{1 - 2^{-d}}.$$

In particular, for a one-dimensional problem ($d = 1$), the storage requirement is less than twice that of the fine-grid problem alone. For problems in two or more dimensions, the requirement drops to less than $\frac{4}{3}$ of the fine-grid problem alone (Exercise 3). Thus, the storage costs of multigrid algorithms decrease relatively as the dimension of the problem increases.

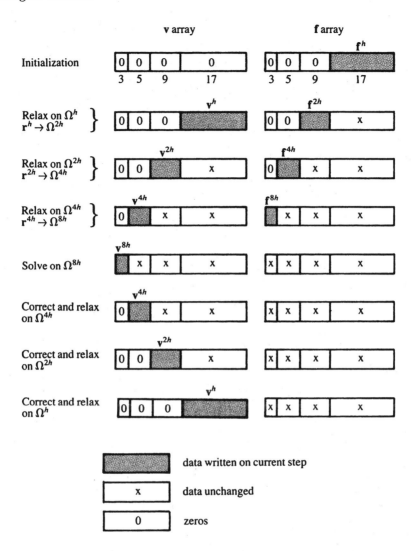

Figure 4.1: *Illustration of the course of a four-level (n = 16) V-cycle showing changes in the data arrays. The v and f arrays hold the solution vectors and right-side vectors, respectively, in the four grids.*

We may use similar reasoning to estimate the computational cost of multigrid methods. It is convenient to measure these costs in terms of a *work unit* (WU), which is the cost of performing one relaxation sweep on the finest grid. It is customary to neglect the cost of intergrid transfer operations, which typically amounts to 10–20% of the cost of the entire cycle.

First consider a V-cycle with one relaxation sweep on each level ($\nu_1 = \nu_2 = 1$). Each level is visited twice and grid Ω^{ph} requires p^{-d} work units. Adding these costs and again using the geometric series for an upper bound gives

V-cycle computation cost

$$= 2\{1 + 2^{-d} + 2^{-2d} + \cdots + 2^{-nd}\} < \frac{2}{1 - 2^{-d}} \text{ WU.}$$

A single V-cycle costs about 4 WUs for a one-dimensional ($d = 1$) problem, about $\frac{8}{3}$ WUs for $d = 2$, and $\frac{16}{7}$ WUs for $d = 3$ (Exercise 4).

With a slight modification, we can find the computational cost for an FMG cycle. Assume again that one relaxation sweep is done on each level ($\nu_0 = \nu_1 = \nu_2 = 1$). As just shown, a full V-cycle beginning from Ω^h costs about $2(1 - 2^{-d})^{-1}$ WUs. A V-cycle beginning from Ω^{2h} costs 2^{-d} of a full V-cycle. In general, a V-cycle beginning from Ω^{ph} costs p^{-d} of a full V-cycle. Adding these costs gives us

> FMG computation cost
$$= \left(\frac{2}{1 - 2^{-d}} \right) \left(1 + 2^{-d} + 2^{-2d} + \cdots + 2^{-nd} \right) < \frac{2}{(1 - 2^{-d})^2} \text{ WU.}$$

An FMG cycle costs 8 WUs for a one-dimensional problem, about $\frac{7}{2}$ WUs for $d = 2$, and $\frac{5}{2}$ WU for $d = 3$ (Exercise 5).

As expected, a single FMG cycle costs more than a single V-cycle, although the discrepancy is less for higher-dimensional problems. We really need to know how many V-cycles and FMG cycles are needed to obtain satisfactory results. This begs the fundamental question: how well do these multigrid cycling schemes work?

Predictive Tools: Local Mode Analysis

The previous section dealt with the practical considerations of implementing multigrid algorithms. However, it is a common experience to have a multigrid code that runs, but does not work! Indeed, it can often be puzzling to know what to expect in terms of efficiency and accuracy. The remainder of this chapter presents some practical tools for determining whether an algorithm is working properly. First, we deal with tools for predicting the convergence rates that can be expected from the basic relaxation methods applied to standard problems.

Recall from Chapter 2 that the asymptotic convergence factor of a relaxation scheme is the spectral radius (the largest eigenvalue magnitude) of the corresponding iteration matrix. We also defined the smoothing factor as the convergence factor associated with the oscillatory modes only. Because eigenvalue calculations can be difficult, this approach to finding convergence factors is limited to fairly simple iterations applied primarily to model problems.

We now present a more versatile approach for approximating convergence and smoothing factors called *local mode analysis* (or *normal mode analysis* or *Fourier analysis*). The goal of this section is rather modest: we show how to apply the basic procedure to some prototype problems and then point the way to more advanced calculations. In its full generality, local mode analysis can be applied to general operators and to a wide class of relaxation schemes on one or more levels. With this generality, local mode analysis is a powerful predictive tool that can be used to compare multigrid performance with theoretical expectations.

The original proponent of local mode analysis was Achi Brandt, who expressed its significance by saying that

> ...the main importance of the smoothing factor is that it separates the design of the interior relaxation from all other algorithmic questions. Moreover, it sets an ideal figure against which the performance of the full algorithm can later be judged. [4]

Local mode analysis begins with the assumption that relaxation is a local process: each unknown is updated using information from nearby neighbors. Because it is a local process, it is argued that boundaries and boundary conditions can be neglected if we are considering a few relaxation sweeps at interior points. For this reason, the finite domain of the problem is replaced by an infinite domain.

As before, we are interested in how a particular relaxation scheme acts on the errors in an approximation. Assume that relaxation is a linear process and denote the associated matrix by R. Let $\mathbf{e}^{(m)}$ denote the algebraic error at the mth step of relaxation. Recall (Chapter 2) that the error itself evolves under the action of R:

$$\mathbf{e}^{(m+1)} = R\mathbf{e}^{(m)}.$$

The approach of local mode analysis is to assume that the error consists of Fourier modes and to determine how relaxation acts on those modes. The Fourier modes we encountered in Chapter 2 have the form $w_j = \sin(\frac{jk\pi}{n})$, where the wavenumber k is an integer between 1 and n. This means that the term $\theta = \frac{k\pi}{n}$ runs roughly from 0 to π. With the new assumption of an infinite domain (no boundaries or boundary conditions to satisfy), the Fourier modes need not be restricted to discrete wavenumbers. Instead, we consider modes of the form $w_j = e^{\iota j\theta}$, where the wavenumber θ can take on any value in the interval $(-\pi, \pi]$. (For the remainder of the chapter, we let $\iota = \sqrt{-1}$ to avoid confusing i with the grid indices.) Notice that the mode corresponding to a particular θ has a wavelength of $\frac{2\pi h}{|\theta|}$; values of $|\theta|$ near zero correspond to low-frequency waves; value of $|\theta|$ near π correspond to high-frequency waves. The choice of a complex exponential makes computations much easier and accounts for both sine and cosine terms.

An important point should be mentioned here. Local mode analysis is not completely rigorous unless the Fourier modes are eigenvectors of the relaxation matrix, which is not generally the case. However, the analysis is useful for the high frequency modes of the error, which *do* tend to resemble the eigenvectors of the relaxation matrix very closely. For this reason, local mode analysis is used for a smoothing analysis of the high frequency modes.

With these ground rules, we are ready to apply the method. We begin with one-dimensional problems and assume that the error at the mth step of relaxation at the jth grid point consists of a single mode of the form

$$e_j^m = A(m)e^{\iota j\theta}, \quad \text{where} \quad -\pi < \theta \le \pi. \tag{4.1}$$

The goal is to determine how the amplitude of the mode, $A(m)$, changes with each relaxation sweep. In each case we consider, the amplitudes at successive steps are related by an expression of the form

$$A(m+1) = G(\theta)A(m).$$

The function G that describes how the error amplitudes evolve is called the *amplification factor*. For convergence of the method, we must have $|G(\theta)| < 1$ for all θ. As we have seen, relaxation is used in multigrid to eliminate the oscillatory modes of the error. Therefore, the quantity of interest is really the *smoothing factor*, which is found by restricting the amplification factor, $G(\theta)$, to the oscillatory modes $\frac{\pi}{2} \le |\theta| \le \pi$. Specifically, we define the smoothing factor as

$$\mu = \max_{\frac{\pi}{2} \le |\theta| \le \pi} |G(\theta)|.$$

This is the factor by which we can expect the oscillatory modes to be damped (at worst) with each relaxation sweep. With these definitions, it is best to proceed by example.

Example: One-dimensional problems. Consider the one-dimensional model problem

$$-u''(x) + c(x)u(x) = f(x).$$

Letting v_j be the approximation to $u(x_j)$, we discretize the problem with the usual second-order finite-difference approximations and apply weighted Jacobi relaxation. This results in the familiar Jacobi updating step

$$v_j^{m+1} = \frac{\omega}{2 + h^2 c_j}(v_{j-1}^m + v_{j+1}^m + h^2 f_j) + (1 - \omega)v_j^m, \tag{4.2}$$

where $c_j = c(x_j)$. Knowing that the error, $e_j = u(x_j) - v_j$, is also governed by the same weighted Jacobi relaxation, we can write the updating step for the error at the jth grid point as (Exercise 6)

$$e_j^{m+1} = \frac{\omega}{2 + h^2 c_j}(e_{j+1}^m + e_{j-1}^m) + (1 - \omega)e_j^m. \tag{4.3}$$

Assume now that the error consists of a mode of the form (4.1) and substitute it into (4.3). Letting $c_j = 0$ for the moment, we have

$$A(m+1)e^{\iota j\theta} = \frac{\omega}{2}\left(A(m)\underbrace{(e^{\iota(j+1)\theta} + e^{\iota(j-1)\theta})}_{2e^{\iota j\theta}\cos\theta}\right) + (1 - \omega)A(m)e^{\iota j\theta}.$$

As indicated, the Euler formula for $\cos\theta$ allows for some simplification. Collecting terms now leads to

$$A(m+1)e^{\iota j\theta} = A(m)(1 - \omega\underbrace{(1 - \cos\theta)}_{2\sin^2(\theta/2)})e^{\iota j\theta}.$$

Canceling the common term $e^{\iota j\theta}$ and using the indicated trigonometric identity, we can write the following relationship between successive amplitudes:

$$A(m+1) = \left(1 - 2\omega\sin^2\left(\frac{\theta}{2}\right)\right)A(m) \equiv G(\theta)A(m), \quad \text{where} \quad -\pi < \theta \le \pi.$$

The amplification factor $G(\theta) = 1 - 2\omega\sin^2(\theta/2)$ appears naturally in this calculation and it should look familiar. In this case, we have just reproduced the eigenvalue calculation for the weighted Jacobi iteration matrix (see Chapter 2); that is, if we make the substitution $\theta_k = \frac{\pi k}{n}$, then $G(\theta_k)$ is just the kth eigenvalue of the Jacobi iteration matrix. As we know, $|G(\theta)| < 1$ provided $0 < \omega \le 1$; with $\omega = \frac{2}{3}$, we have the optimal smoothing factor

$$\mu = G\left(\frac{\pi}{2}\right) = |G(\pm\pi)| = \frac{1}{3}.$$

A similar calculation can be done with Gauss–Seidel. The updating step for the error at the jth grid point now appears as (Exercise 7)

$$e_j^{m+1} = \frac{e_{j-1}^{m+1} + e_{j+1}^m}{2 + c_j h^2}. \tag{4.4}$$

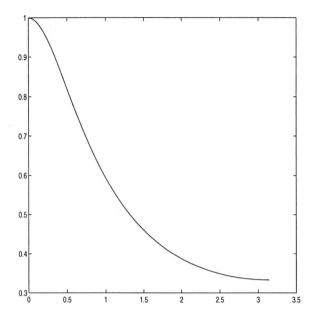

Figure 4.2: *The amplification factor, $|G(\theta)|$, for the Gauss–Seidel method applied to the one-dimensional problem $-u''(x) = f(x)$. The graph is symmetric about $\theta = 0$. The smoothing factor is $\mu = |G(\frac{\pi}{2})| = \frac{1}{\sqrt{5}} = 0.45$.*

Note that because we sweep across the grid from left to right, the previous $((j-1)$st) component has already been updated. Again we assume that the errors have the form (4.1) and substitute. Assuming for the moment that $c_j = 0$, we find that the amplitudes are related by (Exercise 7)

$$A(m+1) = \frac{e^{\iota\theta}}{2 - e^{-\iota\theta}} A(m) \equiv G(\theta)A(m), \quad \text{where} \quad -\pi < \theta \leq \pi.$$

To find the smoothing factor from the complex amplification factor, it is easiest to plot $|G(\theta)|$, as shown in Fig. 4.2. A bit of analysis reveals that

$$\mu = \left| G\left(\frac{\pi}{2}\right) \right| = \frac{1}{\sqrt{5}} = 0.45.$$

A subtle point could be made here. The amplification factor, $G(\theta)$, gives the (complex) eigenvalues of the Gauss–Seidel iteration matrix, not on a bounded domain with specified boundary conditions, but on an infinite domain. This calculation differs from the eigenvalue calculation of Chapter 2, in which the eigenvalues for a bounded domain were found to be real. For this reason, the amplification factor gives only an estimate of the smoothing factors for a bounded domain problem.   ⬦

We can use the above example to illustrate how local mode analysis works with a variable coefficient operator. Suppose that $c(x) > 0$ on the domain. To avoid working with a different amplification factor at every grid point, the practice is to "freeze" the coefficient, $c(x)$, at a representative value $c_0 = c(\xi)$, for some ξ in the domain (often the minimum or maximum value of $c(x)$ on the domain). With the weighted Jacobi iteration, the amplification factor now appears as

$$G(\theta) = 1 - \omega \left(1 - \frac{2}{2 + c_0 h^2} \cos\theta \right).$$

The idea is to find the value of c_0, over all possible $c(\xi)$, that gives the worst (most pessimistic) smoothing factor. Occasionally, this calculation can be done analytically; more typically, it is done numerically by choosing several different possible values of c_0. We can rewrite this amplification factor as

$$G(\theta) = G_0(\theta) - \frac{c_0 \omega h^2}{2} \cos(\theta),$$

where $G_0(\theta)$ is the amplification factor for the case that $c(x) = 0$. In this form, we see that the effect of the variable coefficient is insignificant unless c_0 is comparable to h^{-2}. There is a more general principle at work here: usually the lower order terms of the operator can be neglected with impunity in local mode analysis.

Local mode analysis can be extended easily to two or more dimensions. In two dimensions, the Fourier modes have the form

$$e_{jk}^{(m)} = A(m)e^{\iota(j\theta_1 + k\theta_2)}, \tag{4.5}$$

where $-\pi < \theta_1, \theta_2 \leq \pi$ are the wavenumbers in the x- and y-directions, respectively. Substituting this representation into the error updating step generally leads to an expression for the change in the amplitudes of the form

$$A(m + 1) = G(\theta_1, \theta_2)A(m).$$

The amplification factor now depends on two wavenumbers. The smoothing factor is the maximum magnitude of the amplification factor over the oscillatory modes. As we see in Fig. 4.3, the oscillatory modes correspond to $\frac{\pi}{2} \leq |\theta_i| \leq \pi$ for either $i = 1$ or $i = 2$; that is,

$$\mu = \max_{\pi/2 \leq |\theta_i| \leq \pi} |G(\theta_1, \theta_2)|.$$

Example: Two-dimensional problems. Consider the model problem

$$u_{xx} + u_{yy} = f(x, y)$$

on a rectangular domain with a uniform grid in both directions. Applying the weighted Jacobi method, the error satisfies (Exercise 8)

$$e_{jk}^{(m+1)} = \frac{\omega}{4} \left(e_{j-1,k}^{(m)} + e_{j+1,k}^{(m)} + e_{j,k-1}^{(m)} + e_{j,k+1}^{(m)} \right) + (1 - \omega)e_{jk}^{(m)}. \tag{4.6}$$

Substituting the Fourier modes (4.5) into the error updating equation, we find that (Exercise 8)

$$A(m + 1) = \left[1 - \omega \left(\sin^2\left(\frac{\theta_1}{2}\right) + \sin^2\left(\frac{\theta_2}{2}\right) \right) \right] A(m) \equiv G(\theta_1, \theta_2)A(m).$$

Two views of the amplification factor are given in Fig. 4.4 for the case that $\omega = \frac{4}{5}$. In the left figure, each curve shows the variation of G over $0 \leq \theta_2 \leq \pi$ for fixed

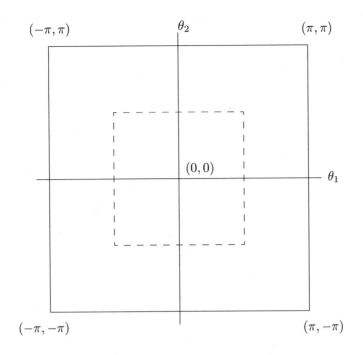

Figure 4.3: *The oscillatory modes in two dimensions correspond to the wavenumbers* $\frac{\pi}{2} \leq |\theta_i| < \pi$ *for either* $i = 1$ *or* $i = 2$; *this is the region outside the dashed box.*

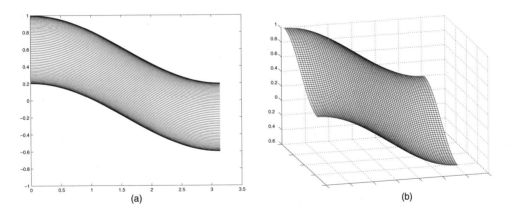

Figure 4.4: (a) *Amplification factor,* $G(\theta_1, \theta_2)$, *for the weighted Jacobi method applied to the model problem in two dimensions, shown as individual curves of fixed* θ_1 ($\theta_1 = 0$ *at the top and* $\theta_1 = \pi$ *at the bottom*). (b) *Same amplification factor shown as a surface over the region* $[0, \pi] \times [0, \pi]$. *The picture is symmetric about both the* θ_1- *and* θ_2-*axes.*

values of θ_1; the upper curve corresponds to $\theta_1 = 0$ and the lower curve corresponds to $\theta_2 = \pi$. Clearly, the amplification factor decreases in magnitude as the modes become more oscillatory. The right figure shows the same amplification factor as a

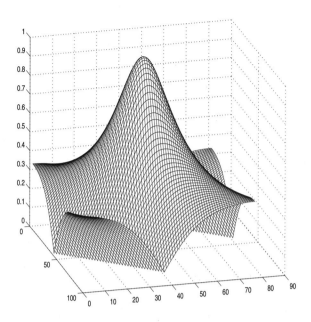

Figure 4.5: *Amplification factor,* $|G(\theta_1, \theta_2)|$*, for the Gauss–Seidel method applied to the model problem in two dimensions, shown as a surface over the region* $[-\pi, -\pi] \times [\pi, \pi]$*.*

surface over the region $[0, \pi] \times [0, \pi]$. (The surface is symmetric about both axes.) Some analysis or experimentation reveals that the best smoothing factor is obtained when $\omega = \frac{4}{5}$ and it is given by $\mu = |G(\pm\pi, \pm\pi)| = 0.6$ (Exercise 8). This means that if we use the Jacobi scheme with $\omega = \frac{4}{5}$, then we expect a reduction in the residual norm by approximately a factor of 0.6 per relaxation sweep. A V(2,1)-cycle, for example, should have a convergence factor of about $0.6^3 = 0.216$.

A similar calculation can be done for the Gauss–Seidel method applied to the model problem. The error updating equation is (Exercise 9)

$$e_{jk}^{(m+1)} = \frac{e_{j-1,k}^{(m+1)} + e_{j+1,k}^{(m)} + e_{j,k-1}^{(m+1)} + e_{j,k+1}^{(m)}}{4}. \tag{4.7}$$

Here we assume that the unknowns have a lexicographic ordering (in order of increasing j and k); thus, the unknowns preceding e_{jk} are updated and appear at the $(m + 1)$st step. Once again, we substitute the modes given in (4.5). The amplification factor is most easily expressed in complex form as (Exercise 9)

$$G(\theta_1, \theta_2) = \frac{e^{\iota\theta_1} + e^{\iota\theta_2}}{4 - e^{-\iota\theta_1} - e^{-\iota\theta_2}}.$$

The magnitude of this function is plotted over the region $[-\pi, -\pi] \times [\pi, \pi]$ in Fig. 4.5.

Some computation is required to show that

$$|G(\theta_1, \theta_2)|^2 = \frac{1 + \cos\beta}{9 - 8\cos(\frac{\alpha}{2})\cos(\frac{\beta}{2}) + \cos\beta},$$

The Discrete L^2 Norm. Another norm that is particularly appropriate for measuring errors in numerical calculations is the *discrete L^2 norm*. If the vector $\mathbf{u}^h$ is associated with a d-dimensional domain with uniform grid spacing h, then its discrete L^2 norm is given by

$$\|\mathbf{u}^h\|_h = \left(h^d \sum_i (u_i^h)^2 \right)^{1/2},$$

which is the usual Euclidean vector norm, scaled by a factor (h^d) that depends on the geometry of the problem. This scaling factor is introduced to make the discrete L^2 norm an approximation to the continuous L^2 norm of a function $u(\mathbf{x})$, which is given by

$$\|u\|_2 = \left(\int_\Omega u(\mathbf{x})d\mathbf{x} \right)^{1/2}.$$

For example, with $d = 1$, let $u(x) = x^{m/2}$, where $m > -1$ is an integer. Also, let $\Omega = [0,1]$ with grid spacing $h = \frac{1}{n}$. Then the associated vector is $u_i^h = x_i^{m/2} = (ih)^{m/2}$. The continuous L^2 norm is

$$\|u\|_2 = \left(\int_0^1 x^{m/2} \right)^{1/2} dx = \frac{1}{\sqrt{m+1}},$$

while the corresponding discrete L^2 norm is

$$\|\mathbf{u}^h\|_h = \left(h \sum_{i=1}^n ((ih)^{m/2})^2 \right)^{1/2} \overset{h \to 0}{=} \frac{1}{\sqrt{m+1}}.$$

In this case, the discrete L^2 norm approaches the continuous L^2 norm as $h \to 0$, which occurs only because of the scaling (see Exercise 18).

where $\alpha = \theta_1 + \theta_2$ and $\beta = \theta_1 - \theta_2$. Restricting the amplification factor to the oscillatory modes, a subtle analysis [25] reveals that the smoothing factor is given by

$$\mu = G\left(\frac{\pi}{2}, \cos^{-1}\left(\frac{4}{5} \right) \right) = \frac{1}{2}.$$

⬦⬦

These examples illustrate local mode analysis for relatively elementary problems. The same technique can be extended, usually with more computation and analysis, to anisotropic equations (for example, $\epsilon u_{xx} + u_{yy} = f$) and line relaxation, as discussed in Chapter 7. It can be used for more general operators (for example, convection-diffusion) and for systems of equations. It can also be applied to other relaxation methods with different orderings, some of which lead to new complications. For example, red-black relaxation has the property that Fourier modes become mixed in pairs (in one dimension) or in groups of four (in two dimensions). Thus, the amplification factor is replaced by an *amplification matrix*. The extension to the coarse-grid correction scheme [10] on two levels requires an analysis of interpolation and restriction in a Fourier setting, a subject discussed in the next chapter.

Diagnostic Tools

As with any numerical code, debugging can be the most difficult part of creating a successful program. For multigrid, this situation is exacerbated in two ways. First, the interactions between the various multigrid components are very subtle, and it can be difficult to determine which part of a code is defective. Even more insidious is the fact that an incorrectly implemented multigrid code can perform quite well—sometimes better than other solution methods! It is not uncommon for the beginning multigrid user to write a code that exhibits convergence factors in the 0.2–0.3 range for model problems, while proper tuning would improve the factors to something more like 0.05. The difficulty is convincing the user that 0.2–0.3 is not good enough. After all, this kind of performance solves the problem in very few cycles. But the danger is that performance that is below par for model problems might really expose itself as more complexities are introduced. Diagnostic tools can be used to detect defects in a multigrid code—or increase confidence in the observed results.

Achi Brandt has said that "the amount of computational work should be proportional to the amount of real physical changes in the computed system" and "stalling numerical processes must be wrong." These statements challenge us to develop codes that achieve the best possible multigrid performance. The following short list of diagnostic tools should help to achieve that goal. A systematic approach to writing multigrid codes is also given by Brandt in the section "Stages in Developing Fast Solvers" in his 1984 *Guide* [4].

Of course, no short list of debugging techniques can begin to cover all contingencies. What we provide here is a limited list of tricks and techniques that can be useful in evaluating a multigrid code; they often tell more about the *symptoms* of a defective code than the *causes*. Nevertheless, with increasing experience, they can guide the user to develop properly tuned codes that achieve multigrid's full potential.

- **Methodical Plan.** The process of testing and debugging a multigrid code should be planned and executed methodically. The code should be built in a modular way so that each component can be tested and integrated into the evolving code with confidence. It is best to test the algebraic solver first (for example, V-cycles); then the discretization can be tested, followed by the FMG solver, if it is to be used. In other words, the initial focus should be on ensuring that the basic cycling process solves the discrete system up to expectations. This solver can then be used to test discretization accuracy. Poor discretization, especially at boundaries, is often the source of multigrid inefficiency. Therefore, it is important to test the discretization, perhaps with another solver, when multigrid troubles persist. FMG requires an efficient V-cycle or W-cycle solver *and* an accurate discretization method. This means that FMG should be implemented and tested after the solver and the discretization components are verified (see last item).

- **Starting Simply.** This recommendation is perhaps the most obvious: it is always best to begin with basic methods applied to small, simple problems. Simpler cases expose troubles more clearly and make it easier to trace the sources. Test problems should consist of either a discrete system with a known solution or the simplest form of the desired PDE (usually this means constant coefficients, no convection, no nonlinearity, and trivial boundary conditions). A good sequence is to test the solver on the coarsest grid that your code accepts, then add one finer level to test the two-grid scheme thoroughly. One can then proceed progressively and methodically to larger problems. Once the simplest cases show expected performance, complexities can be added one at a time.

- **Exposing Trouble.** It is critical to start with simple problems so that potential difficulties are kept in reserve. At the same time, it is also important to avoid aspects of the problem that mask troubles. For example, reaction terms can produce strong enough diagonal dominance in the matrix that relaxation by itself is efficient. These terms should be eliminated in the initial tests if possible. Similarly, if the matrix arises from implicit time-stepping applied to a time-dependent problem, then a very large or even infinite time step should be taken in the initial tests.

- **Fixed Point Property.** Relaxation should not alter the exact solution to the linear system: it should be a fixed point of the iteration. Thus, using the exact solution as an initial guess should yield a zero residual before *and* after the relaxation process. Furthermore, the coarse-grid problem takes the transferred residual as its right side, which means that its solution should also be zero. Because the coarse-grid problem uses zero as the initial guess, it is solved exactly, and the correction passed up to the finer grid is also zero. Therefore, neither relaxation nor coarse-grid correction should alter the exact solution.

  This property can be checked by creating a right-side vector corresponding to the known solution (of the linear system) and then using that solution as an initial guess. The relaxation module should be tested first, after which the V-cycle can be tested. If the output from either differs by more than machine precision from the input, there must be an error in the code.

- **Homogeneous Problem.** Applying a multigrid V-cycle code to a homogeneous problem has the advantage that both the norm of the residual and the norm of the error are computable and should decrease to zero (up to machine precision) at a steady rate; it may take eight to ten V-cycles for the steady rate to appear. The predictive mode analysis tools described above can be used to determine the factor by which the residual norm should decrease; it should tend to the asymptotic factor predicted by the smoothing analysis.

- **Zero Residuals.** A useful technique is to multiply the residual by zero just prior to transferring it to the coarse grid. As in the homogeneous problem, the coarse-grid problem now has a zero right side, so its solution is zero. Because the initial guess to the coarse-grid problem is zero, it is solved exactly, and the correction (also zero) is passed back to the fine grid. The utility of this test is that the only part of the code now affecting the approximation is relaxation on the fine grid. This means that the sequence of approximations generated by $V(\nu_1, \nu_2)$-cycles should be identical to the approximations generated using only relaxation; this is an easy property to test.

- **Residual Printout.** A good technique for monitoring performance is to print out the norm of the residual on each level after relaxation on both the descent and ascent legs of the V-cycle. Here again, it is important that the discrete L^2 norm be used, because its scaling makes the error norms comparable from one level to the next. The residuals should behave in a regular fashion: on each level, the sequence of norms should decline to machine zero at a steady rate as the cycling continues. The norm on each level should be smaller after post-relaxation (on the upward leg) than it was after pre-relaxation (on the downward leg). This ensures that the coarse-grid correction/relaxation tandem is working on each level.

  Because most multigrid codes are recursive in nature (although they may not actually use recursive calls), it is important to note that seeing an abnormal residual pattern on a given level does not necessarily mean the code is somehow wrong on that level. More frequently, the flaw exists on all levels, because all levels are treated by the same code. Indeed, the most common culprits are the intergrid transfer operators and boundary treatments. Monitoring the sequence of residuals, however, can be more helpful in ascertaining the presence of a problem than simply observing the overall convergence rate on the fine grid.

- **Error Graph.** Solver trouble that seems impervious to diagnosis can sometimes be resolved with a picture of the error. Surface plots of the algebraic error before and after relaxation on the fine grid can be extremely informative. Of course, knowledge of the error is required, so solving the homogeneous problem can be advantageous here. Is the error oscillatory after coarse-grid correction? Is it effectively smoothed by relaxation everywhere? Is there unusual behavior of the error near special features of the domain such as boundaries or interfaces?

- **Two-Level Cycles.** For any multigrid method to work, it is necessary that the two-level scheme (relaxation and exact coarse-grid correction) work. A useful technique is to test the two-level scheme. This may be done by replacing the recursive call for V-cycles with a direct or iterative solver on the coarse grid. If an iterative solver is used (many multigrid cycles on the coarse grid might actually be used here), the coarse-grid problem should be solved very accurately, possibly to machine precision. The correction can then be transferred to the fine grid and applied there.

  Another useful trick is to use two-level cycling between specific pairs of levels. In particular, if the residual printouts indicate an abnormal residual on a specific level, it is useful to perform two-level cycling between that level and the level below it (or between that level and the one above it). This may isolate exactly where the problem occurs.

- **Boundaries.** One of the most common sources of errors in multigrid programs is the incorrect treatment of boundary data. Generally, interpolation and restriction stencils must be altered at boundary points and neighbors of boundary points. Issues involved here are often very subtle and errors can be difficult to detect. Among the most common indicators is that V-cycles will show convergence factors at or near the predicted values for several V-cycles, only to have convergence slow down and eventually stall with continued cycling. The reason is that early in the process, errors at the boundaries have little effect, since the boundary is a lower-dimensional feature of the problem. However, as the cycling continues, the errors from the boundaries propagate into the interior, eventually infecting the entire domain.

  One of the most useful techniques in building or debugging a multigrid code is to separate the effects of the boundary entirely. This can be done by replacing the given boundary conditions by periodic conditions. A periodic boundary value problem should attain convergence factors near those predicted by mode analysis, because the assumptions underlying the Fourier analysis are more closely met. Once it is determined that periodic boundary conditions work properly, the actual boundary conditions may be applied. If the convergence factors degrade with continued cycling, then the treatment at the boundaries may be suspected. A useful technique is to apply extra relaxation sweeps at the boundaries. Often, the extra relaxations will overcome any boundary difficulties and the overall results will approach the ideal convergence factors. If this does not occur, then a very careful examination of how each piece of the code treats the boundaries is in order. Special attention should be paid to the intergrid transfer operators at and near the boundaries. Just as it is easy to obtain discretizations that are of lower order on the boundary than the interior, it is also easy to produce intergrid transfers that fail to be consistent in their treatment of boundary data.

- **Symmetry.** Matrices that arise in the discretization of self-adjoint problems are often symmetric. Such is the case for most of the matrices featured in this book. Many coarsening strategies preserve matrix symmetry when it is present, as does the so-called Galerkin scheme introduced in the next chapter. Inadvertent loss of symmetry often occurs at boundaries, especially at corners and other irregularities of the domain. This loss can lead to subtle difficulties that are hard to trace. If the fine- and coarse-grid matrices are supposed to be symmetric, then this should be tested. This is easily done by comparing entries i, j and j, i of the matrix on each grid.

- **Compatibility Conditions.** We have not yet discussed compatibility conditions, but they arise in the context of Neumann boundary conditions (among other settings), which we discuss in Chapter 7. A compatibility condition is one that must be enforced on the right-side vector to ensure that the problem is solvable. A common source of error is that, while great care may be taken to enforce a compatibility condition on the fine grid, it might not be enforced on the coarse grids. Sometimes, the compatibility condition is enforced automatically in the coarsening process. However, round-off errors may enter and compound themselves as coarsening proceeds, so it may be worthwhile to enforce the condition explicitly on all coarse grids.

- **Linear Performance by Nonlinear Codes.** We describe the FAS (Full Approximation Scheme) for nonlinear problems in Chapter 6. A useful technique for debugging an FAS code is to begin by applying it to a linear problem. FAS reduces to standard multigrid in this case, so an FAS code should exhibit standard performance on these problems. FAS codes should be written so the nonlinearity can be controlled by a parameter: setting the parameter to zero yields a linear problem, while increasing the parameter value strengthens the nonlinearity. As the nonlinearity increases, the performance of FAS should not, in general, degrade significantly (at least for reasonable values of the parameter).

- **Solutions to the PDE.** A known solution to the underlying PDE can be very useful in assessing whether a multigrid code is working as it should. The first thing to consider is whether the solution computed from the multigrid code looks like a sampled version of the continuous solution. The first cursory examination can be qualitative: Does the overall shape of the computed solution resemble the exact solution? Are the peaks and valleys in the right places?

  If the qualitative comparison is good, more quantitative tests can be performed. First, the norm of the error (the difference between the sampled continuous solution and the approximation) should be monitored as a function of the number of V-cycles. The discrete L^2 norm is usually appropriate here. This norm should behave in very specific ways, depending on the accuracy of the discretization. If the known solution has no discretization error (for example, if second-order finite differences are used and the known solution is a second degree polynomial), then the error norm should be driven steadily to machine zero with continued V-cycles. Indeed, the rate at which it goes to zero should be about the same rate at which the norm of the residual declines, and it should reflect the predicted asymptotic convergence factor.

  On the other hand, if discretization error is present, then we expect the norm to stop decreasing after several V-cycles, and it may even grow slightly. This indicates (in a correctly working code) that we have reached the *level of discretization error* (roughly the difference between the continuous and discrete solutions, as discussed in Chapter 5). Here it is useful to solve the same problem repeatedly, with the same right side, on a sequence of grids with decreasing grid spacing h. If the discretization is, for example, $O(h^2)$ accurate in the discrete L^2 norm, then the error norms should decrease roughly by a factor of four each time h is halved. Naturally, the fit will not be perfect, but if the code is working correctly, the trend should be very apparent.

- **FMG Accuracy.** The basic components of an effective FMG scheme are an efficient V-cycle (or W-cycle) solver and an accurate discretization. The idea is to assess the discretization error for a given problem using a sequence of increasingly finer grids. This can be done by choosing a PDE with a known solution and solving each level in turn, starting from the coarsest. Each level should be solved very accurately, perhaps using many V-cycles, testing the residual norm to be sure it is small. The computed solution can then be compared on each level to the PDE solution, evaluated at the grid points, using the discrete L^2 norm. These comparisons yield discretization error estimates on each level. FMG can then be tested by comparing its computed approximation on each level to the PDE solution. The ratio of these estimates to the discretization error estimates should be close to one as the mesh size decreases. This signals that FMG is achieving accuracy to the level of discretization error. Properly tuning the FMG scheme by choosing the right number of V-cycles, pre-smoothing sweeps, and post-smoothing sweeps may be required to achieve this property.

Having discussed a variety of practical matters, it is now time to observe these ideas at work in some numerical experiments.

Numerical example. The first experiment deals with the one-dimensional model problem $A\mathbf{u} = \mathbf{0}$. The weighted Jacobi method with $\omega = \frac{2}{3}$ is applied to this problem on a grid with $n = 64$ points. The initial guess consists of two waves with wavenumbers $k = 3$ and $k = 10$. For purposes of illustration, we implemented a modified V-cycle algorithm. Called the *immediate replacement algorithm*, this version makes an error correction directly to the fine grid after every coarse-grid relaxation. In this way, it is possible to see the immediate effect of each coarse-grid correction on the full, fine-grid solution. Although this version is algebraically equivalent to the standard algorithm, it is impractical because it involves an inordinate number of intergrid transfers. Nevertheless, it is useful for demonstrating algorithm performance because it allows us to monitor the effects of individual coarse-grid operations on the error.

Figures 4.6(a, b) show the course of the algorithm by plotting the maximum norms of the error and the residual after each coarse-grid correction. The algorithm progresses from left to right in the direction of increasing work units (WUs). Notice that the data points are spaced nonuniformly due to the different amount of work done on each level.

Figure 4.6(a) illustrates five full V-cycles with $\nu_1 = \nu_2 = 1$ relaxation sweep on each grid. This computation requires roughly 15 WUs. In these 15 WUs, the initial error norm is reduced by about three orders of magnitude, giving an average convergence rate of about 0.2 decimal digits per WU. Figure 4.6(b) shows the result of performing V-cycles with $\nu_1 = \nu_2 = 2$ relaxation sweeps on each level. In this case, three V-cycles are done in 20 WUs, giving an average convergence rate of about 0.15 decimal digits per WU. In this case, it seems more efficient to relax only once on each level. For another problem, a different cycling strategy might be preferable. It is also possible to relax, say, twice on the "descent" phase of the V-cycle and only once on the "ascent" phase.

The curves of these two figures also show a lot of regular fine structure. Neither the error norm nor the residual norm decreases monotonically. The error curve decreases rapidly on the "descent" to the coarsest grid, while the residual curve decreases rapidly on the "ascent" to the finest grid. A detailed account of this fine structure should be left to multigrid aficionados!   ⋄⋄

The previous experiment indicates that the choice of cycling parameters (ν_0, ν_1, ν_2) for multigrid schemes may not be obvious. They are often chosen *a priori*, based on analysis or prior experimentation, and they remain fixed throughout the course of the algorithm. For certain problems, there are heuristic methods for changing the parameters adaptively [4].

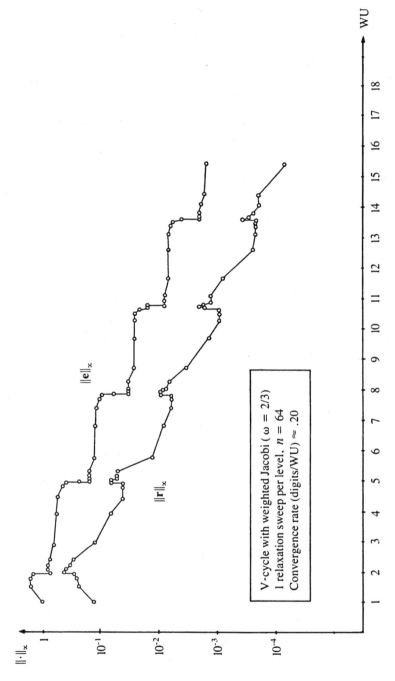

Figure 4.6: (a). *Immediate replacement version of the V-cycle scheme applied to the one-dimensional model problem with $n = 64$ points with $\nu_1 = \nu_2 = 1$ relaxation sweep on each level. The maximum norms of the error and the residual are plotted against WUs over the course of five V-cycles.*

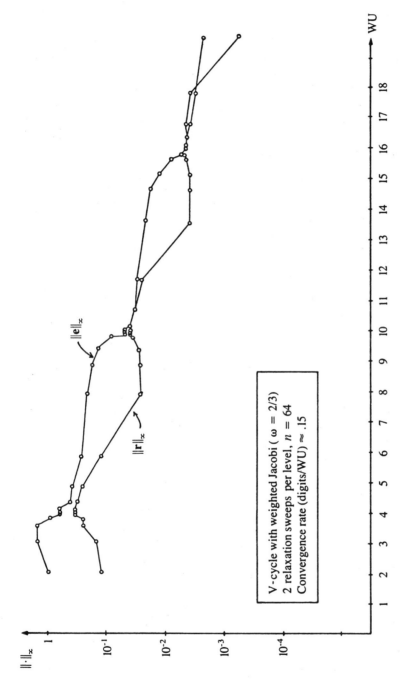

Figure 4.6, continued: (b). *Immediate replacement version of the V-cycle scheme applied to the one-dimensional model problem with $n = 64$ points with $\nu_1 = \nu_2 = 2$ relaxation sweeps on each level. The maximum norms of the error and the residual are plotted against WUs over the course of three V-cycles.*

We have devoted much attention to the one-dimensional model problem with the understanding that many of the algorithms, ideas, and results extend directly to higher dimensions. It is useful to mention a few issues that arise only in higher-dimensional problems. For example, the basic relaxation schemes have many more variations. In two dimensions, we can relax by points (updating one unknown at a time in one-dimensional problems) or by lines. In line relaxation, an entire row or column of the grid is updated at once, which generally requires the direct solution of a tridiagonal system for each line. Line relaxation permits additional options, determined by orderings. The lines may be swept forward, backward, or both (symmetric relaxation). The lines may be colored and relaxed alternately in a red-black fashion (often called *zebra relaxation*). Or the grid may be swept alternately by rows and then by columns, giving an *alternating direction* method. Line relaxation will be discussed further in Chapter 7.

These ideas are further generalized in three dimensions. Here we can relax by points, lines, or planes, with various choices for ordering, coloring, and direction. There are also ways to incorporate fast direct solvers for highly structured subproblems that may be imbedded within a relaxation sweep. Many of these possibilities have been analyzed; many more have been implemented. However, there is still room for more work and understanding.

Numerical example. We conclude this chapter with an extensive numerical experiment in which several multigrid methods are applied to the two-dimensional problem

$$-u_{xx} - u_{yy} = 2\big[(1 - 6x^2)y^2(1 - y^2) + (1 - 6y^2)x^2(1 - x^2)\big] \quad \text{in} \quad \Omega, \quad (4.8)$$
$$u = 0 \qquad\qquad \text{on} \quad \partial\Omega,$$

where Ω is the unit square, $\{(x, y) : 0 < x < 1, \ 0 < y < 1\}$. Knowing that the analytical solution to this problem is

$$u(x, y) = (x^2 - x^4)(y^4 - y^2),$$

errors can be computed.

It should be mentioned in passing that the convergence properties of the basic relaxation methods carry over directly from one to two dimensions when they are applied to the model problem. Most importantly, weighted Jacobi and Gauss–Seidel retain the property that they smooth high-frequency Fourier modes effectively and leave low-frequency modes relatively unchanged. A guided eigenvalue calculation that leads to these conclusions is given in Exercise 12.

We first use red-black Gauss–Seidel relaxation in a V-cycle scheme on fine grids with $n = 16$, 32, 64, and 128 points in each direction (four separate experiments). Full weighting and linear interpolation are used. We let $\mathbf{e}$ be the vector with components $u(x_i) - v_i^h$ and compute its discrete L^2 norm of the error, $\|\cdot\|_h$. Because the error is not available in most problems, a more practical measure, the discrete L^2 norm of the residual $\mathbf{r}^h$, is also computed.

Table 4.1 shows the residual and error norms after each V-cycle. For each V-cycle, the two columns labeled *ratio* show the ratios of $\|\mathbf{r}^h\|_h$ and $\|\mathbf{e}\|_h$ between successive V-cycles. There are several points of interest. First consider the column

V-cycle	$n = 16$				$n = 32$			
	$\|\mathbf{r}^h\|_h$	ratio	$\|\mathbf{e}\|_h$	ratio	$\|\mathbf{r}^h\|_h$	ratio	$\|\mathbf{e}\|_h$	ratio
0	6.75e+02		5.45e−01		2.60e+03		5.61e−01	
1	4.01e+00	0.01	1.05e−02	0.02	1.97e+01	0.01	1.38e−02	0.02
2	1.11e−01	0.03	4.10e−04	0.04	5.32e−01	0.03	6.32e−04	0.05
3	3.96e−03	0.04	1.05e−04	0.26	2.06e−02	0.04	4.41e−05	0.07
4	1.63e−04	0.04	1.03e−04	0.98*	9.79e−04	0.05	2.59e−05	0.59
5	7.45e−06	0.05	1.03e−04	1.00*	5.20e−05	0.05	2.58e−05	1.00*
6	3.75e−07	0.05	1.03e−04	1.00*	2.96e−06	0.06	2.58e−05	1.00*
7	2.08e−08	0.06	1.03e−04	1.00*	1.77e−07	0.06	2.58e−05	1.00*
8	1.24e−09	0.06	1.03e−04	1.00*	1.10e−08	0.06	2.58e−05	1.00*
9	7.74e−11	0.06	1.03e−04	1.00*	7.16e−10	0.06	2.58e−05	1.00*
10	4.99e−12	0.06	1.03e−04	1.00*	4.79e−11	0.07	2.58e−05	1.00*
11	3.27e−13	0.07	1.03e−04	1.00*	3.29e−12	0.07	2.58e−05	1.00*
12	2.18e−14	0.07	1.03e−04	1.00*	2.31e−13	0.07	2.58e−05	1.00*
13	2.33e−15	0.11	1.03e−04	1.00*	1.80e−14	0.08	2.58e−05	1.00*
14	1.04e−15	0.45	1.03e−04	1.00*	6.47e−15	0.36	2.58e−05	1.00*
15	6.61e−16	0.63	1.03e−04	1.00*	5.11e−15	0.79	2.58e−05	1.00*

V-cycle	$n = 64$				$n = 128$			
	$\|\mathbf{r}^h\|_h$	ratio	$\|\mathbf{e}\|_h$	ratio	$\|\mathbf{r}^h\|_h$	ratio	$\|\mathbf{e}\|_h$	ratio
0	1.06e+04		5.72e−01		4.16e+04		5.74e−01	
1	7.56e+01	0.01	1.39e−02	0.02	2.97e+02	0.01	1.39e−02	0.02
2	2.07e+00	0.03	6.87e−04	0.05	8.25e+00	0.03	6.92e−04	0.05
3	8.30e−02	0.04	4.21e−05	0.06	3.37e−01	0.04	4.22e−05	0.06
4	4.10e−03	0.05	7.05e−06	0.17	1.65e−02	0.05	3.28e−06	0.08
5	2.29e−04	0.06	6.45e−06	0.91*	8.99e−04	0.05	1.63e−06	0.50
6	1.39e−05	0.06	6.44e−06	1.00*	5.29e−05	0.06	1.61e−06	0.99*
7	8.92e−07	0.06	6.44e−06	1.00*	3.29e−06	0.06	1.61e−06	1.00*
8	5.97e−08	0.07	6.44e−06	1.00*	2.14e−07	0.06	1.61e−06	1.00*
9	4.10e−09	0.07	6.44e−06	1.00*	1.43e−08	0.07	1.61e−06	1.00*
10	2.87e−10	0.07	6.44e−06	1.00*	9.82e−10	0.07	1.61e−06	1.00*
11	2.04e−11	0.07	6.44e−06	1.00*	6.84e−11	0.07	1.61e−06	1.00*
12	1.46e−12	0.07	6.44e−06	1.00*	4.83e−12	0.07	1.61e−06	1.00*
13	1.08e−13	0.07	6.44e−06	1.00*	3.64e−13	0.08	1.61e−06	1.00*
14	2.60e−14	0.24	6.44e−06	1.00*	1.03e−13	0.28	1.61e−06	1.00*
15	2.30e−14	0.88	6.44e−06	1.00*	9.19e−14	0.89	1.61e−06	1.00*

Table 4.1: *The V(2,1) scheme with red-black Gauss–Seidel applied to a two-dimensional problem on fine grids with $n = 16$, 32, 64, and 128 points. The discrete L^2 norms of the residual and error are shown after each V-cycle. The* ratio *columns give the ratios of residual and error norms of successive V-cycles. The* * *in the error ratio column indicates that the level of discretization error has been reached.*

of error norms. For each of the four grid sizes, the error norms decrease rapidly and then level off abruptly as the scheme reaches the level of discretization error. We confirm this by comparing the final error norms, $\|e\|_h$, on the four grids (1.03e − 04, 2.58e − 05, 6.44e − 06, and 1.61e − 06). These norms decrease by a factor of four as the resolution doubles, which is consistent with the second-order discretization we have used. The residual norms also decrease rapidly for 12 to 14 V-cycles, with the value in the corresponding *ratio* column reaching a nearly constant value, until the last few cycles. This constant value is a good estimate of the asymptotic convergence factor of the scheme (approximately 0.07) for this problem. The sharp increase in the residual norm ratio during the last two cycles reflects the fact that the algebraic approximation is already accurate to near machine precision.

In the course of our development, we described several different schemes for relaxation, restriction, and interpolation. Specifically, we worked with weighted Jacobi, Gauss–Seidel, and red-black Gauss–Seidel relaxation schemes; injection and full weighting restriction operators; and linear interpolation. We now investigate how various combinations of these schemes perform when used in V-schemes applied to model problem (4.8).

For completeness, we introduce two more schemes, *half-injection* and *cubic interpolation*. Half-injection, as the name implies, is simply defined in one dimension by $v_j^{2h} = 0.5v_{2j}^h$, with a similar extension to two dimensions. Half-injection is designed for use on the model problem with red-black relaxation and may be understood most easily by considering the one-dimensional case. The idea is that because one sweep of the red-black relaxation produces zero residuals at every other point, full-weighting and half-injection are equivalent. We will see shortly that the scheme indeed works well for this case.

Cubic interpolation is one of the many interpolation schemes that could be applied and is probably the most commonly used in multigrid after linear interpolation. As its name implies, the method interpolates cubic (or lower degree) polynomials exactly. It represents a good compromise between the desire for greater interpolation accuracy and the increase in computational cost required to achieve the desired accuracy. In one dimension, the basic cubic interpolation operator is given by

$$
\begin{aligned}
v_{2j}^h &= v_j^{2h}, \\
v_{2j+1}^h &= \frac{1}{16}\left(-v_{j-1}^{2h} + 9v_j^{2h} + 9v_{j+1}^{2h} - v_{j+2}^{2h}\right).
\end{aligned}
$$

Table 4.2 gives comparative results for many experiments. For each of the Jacobi, Gauss–Seidel, and red-black Gauss–Seidel relaxation schemes, we performed six V-cycles, using all possible combinations of the half-injection, injection, and full weighting restriction operators with linear or cubic interpolations. In each case, the experiment was performed using (1,0), (1,1), and (2,1) V-cycles, where (ν_1, ν_2) indicates ν_1 pre-correction relaxation sweeps and ν_2 post-correction relaxation sweeps. The entries in the table give the average convergence factor for the last five V-cycles of each experiment. A dash indicates that a particular scheme diverged. The entries in the *cost* line reflect the cost of the method, in terms of operation count, shown as a multiple of the cost of the (1,0) scheme using linear interpolation and injection. Notice that the cost is independent of the different relaxation schemes, as they all perform the same number of operations.

	Relaxation	Injection		Full Weighting		Half-Injection	
(ν_1, ν_2)	Scheme	Linear	Cubic	Linear	Cubic	Linear	Cubic
(1,0)	Jacobi	–	–	0.49	0.49	0.55	0.62
	GS	0.89	0.66	0.33	0.34	0.38	0.37
	RBGS	–	–	0.21	0.23	0.45	0.42
	Cost	1.00	1.25	1.13	1.39	1.01	1.26
(1,1)	Jacobi	0.94	0.56	0.35	0.34	0.54	0.52
	GS	0.16	0.16	0.14	0.14	0.45	0.43
	RBGS	–	–	0.06	0.05	0.12	0.16
	Cost	1.49	1.75	1.63	1.88	1.51	1.76
(2,1)	Jacobi	0.46	0.31	0.24	0.24	0.46	0.45
	GS	0.07	0.07	0.08	0.07	0.40	0.39
	RBGS	–	–	0.04	0.03	0.03	0.07
	Cost	1.99	2.24	2.12	3.37	1.51	1.76

Table 4.2: *Average convergence factors over five V-cycles on model problem* (4.8) *for various combinations of relaxation, restriction, and interpolation operators. The dashes indicate divergent schemes. The* cost *line gives the computational cost of a* V(ν_1, ν_2)*-cycle scheme using the specified choice of restriction and interpolation, as a multiple of the cost of a* (1,0) V*-cycle scheme using injection and linear interpolation.*

A few observations are in order. At least for this problem, cubic interpolation is only noticeably more effective than linear interpolation when injection is used as the restriction operator. It is also apparent that half-injection is useful only with red-black Gauss–Seidel, as expected; even then, several smoothing sweeps are required. Finally, and not surprisingly, the table indicates that you get what you pay for: combinations that produce the best convergence factors are also those with the higher costs. Parameter selection is largely the art of finding a compromise between performance and cost.

The final set of experiments concerns the effectiveness of FMG schemes and, in particular, whether FMG schemes are more or less effective than V-cycles. Table 4.3 describes the performance of three FMG schemes. Square grids with up to $n = 2048$ points in each direction are used. The FMG(ν_1, ν_2) scheme uses ν_1 relaxation sweeps on the descent phase and ν_2 relaxation sweeps on the ascent phase of each V-cycle. Red-black Gauss–Seidel relaxation is used with full weighting and linear interpolation. In each case, only one complete FMG cycle is done.

The table shows the discrete L^2 norm of the error for each scheme. Evidence that the FMG code solves the problem to the level of discretization error on each grid is that the ratio of the error norms between successive levels is roughly 0.25 (for this two-dimensional problem). If the ratio is noticeably greater than 0.25, the solution is probably not accurate to the level of discretization. Based on this observation, we suspect that the FMG(1,0) scheme does not solve the problem to the level of discretization error on any level. This is confirmed when we observe the FMG(1,1) and FMG(2,1) schemes, which *do* solve to the level of the discretization error on all levels. The FMG(2,1) scheme requires more work than the FMG(1,1) with little gain in accuracy; so it appears that FMG(1,1) is the best choice for this problem.

N	FMG(1,0)		FMG(1,1)		FMG(2,1)		FMG(1,1)	V(2,1)	V(2,1)
	$\|e\|_h$	ratio	$\|e\|_h$	ratio	$\|e\|_h$	ratio	WU	cycles	WU
2	5.86e−03		5.86e−03		5.86e−03				
4	5.37e−03	0.917	2.49e−03	0.424	2.03e−03	0.347	7/2	3	12
8	2.78e−03	0.518	9.12e−04	0.367	6.68e−04	0.328	7/2	4	16
16	1.19e−03	0.427	2.52e−04	0.277	1.72e−04	0.257	7/2	4	16
32	4.70e−04	0.395	6.00e−05	0.238	4.00e−05	0.233	7/2	5	20
64	1.77e−04	0.377	1.36e−05	0.227	9.36e−06	0.234	7/2	5	20
128	6.49e−05	0.366	3.12e−06	0.229	2.26e−06	0.241	7/2	6	24
256	2.33e−05	0.359	7.35e−07	0.235	5.56e−07	0.246	7/2	7	28
512	8.26e−06	0.354	1.77e−07	0.241	1.38e−07	0.248	7/2	7	28
1024	2.90e−06	0.352	4.35e−08	0.245	3.44e−08	0.249	7/2	8	32
2048	1.02e−06	0.351	1.08e−08	0.247	8.59e−09	0.250	7/2	9	36

Table 4.3: *Three different FMG schemes applied to the two-dimensional problem on square grids with up to $n = 2048$ points in each direction. The $FMG(\nu_1, \nu_2)$ scheme uses ν_1 red-black Gauss–Seidel relaxation sweeps on the descent phase and ν_2 relaxation sweeps on the ascent phase of each V-cycle. The discrete L^2 norm of the error and the ratio of errors at each grid level are shown. Solution to the level of discretization error is indicated when the ratio column shows a reduction of at least 0.25 in the error norm. For comparison, the V-cycles column shows the number of V(2,1)-cycles needed to converge to the level of discretization error, while the V-cycle WU column shows the number of work units needed to converge to the level of discretization error.*

A useful question is whether an FMG(1,1) scheme is more efficient than, say, the V(2,1) scheme in achieving a solution accurate to the level of discretization error. We answer this question by performing the V(2,1) method (as in Table 4.1) for all grid sizes from $n = 4$ through $n = 2048$ (over 4 million fine-grid points!) and recording the number of V-cycles required to converge to the level of discretization error. These results are presented in the second-to-last column of Table 4.3. It is apparent that the number of cycles required to solve to the level of discretization error increases with the problem size.

We can now make some comparisons. Recall our discussion of computational costs earlier in the chapter. We determined that a (1,1) V-cycle in $d = 2$ dimensions costs about $\frac{8}{3}$ WUs (Exercise 3); therefore, a (2,1) V-cycle costs half again as much, or 4 WU. The last column of Table 4.3 shows the costs in WUs of solving to the level of discretization error with the V(2,1) scheme on various grids. We also saw (Exercise 4) that the FMG(1,0) scheme, which did not converge to the level of discretization error in this case, requires just under 2 WUs, while the FMG(1,1) and FMG(2,1) schemes, which did achieve the desired accuracy, require approximately $\frac{7}{2}$ and $\frac{16}{3}$ WUs, respectively; these costs are the same for all grid sizes. Thus, on all of the grids shown in Table 4.3, the FMG(1,1) scheme is significantly less expensive in WUs than the V(2,1) scheme. This confirms the observation that for converging to the level of discretization error, full multigrid methods are generally preferable to simple V-cycles. ◇◇

Exercises

Data Structures and Complexity

1. **Data structures.** Work out the details of the data structures given in Fig. 4.1. Assume that for a one-dimensional problem, the finest grid has $n - 1 =$

$2^L - 1$ interior points. Let $h = \frac{1}{n}$ be the grid spacing on Ω^h. Let level l have grid spacing $2^{l-1}h$. As suggested in the text, store the approximations $\mathbf{v}^h, \mathbf{v}^{2h}, \ldots$ contiguously in a single array $\mathbf{v}$, with the level L values stored in v_1, v_2, v_3; the level $L-1$ values in $v_4, \ldots, v_8$; etc. Use a similar arrangement for the right-side values $\mathbf{f}^h, \mathbf{f}^{2h}, \ldots$. How many values are stored on level l, where $1 \le l \le L$? What is the starting index in the $\mathbf{v}$ array for the level l values, where $1 \le l \le L$?

2. **Data structure for two dimensions.** Now consider the two-dimensional model problem. The one-dimensional data structure may be retained in the main program. However, the initial index for each grid will now be different. Compute these indices, assuming that on the finest grid Ω^h there are $(n-1)^2$ interior points, where $n - 1 = 2^L - 1$.

3. **Storage requirements.** Verify the statement in the text that for a one-dimensional problem $(d = 1)$, the storage requirement is less than twice that of the fine-grid problem alone. Show that for problems in two or more dimensions, the requirement drops to less than $\frac{4}{3}$ of the fine-grid problem alone.

4. **V-cycle computation cost.** Verify the statement in the text that a single V-cycle costs about 4 WUs for a one-dimensional $(d = 1)$ problem, about $\frac{8}{3}$ WUs for $d = 2$, and $\frac{16}{7}$ WUs for $d = 3$.

5. **FMG computation cost.** Verify the statement in the text that an FMG cycle costs 8 WUs for a one-dimensional problem; the cost is about $\frac{7}{2}$ WUs for $d = 2$ and $\frac{5}{2}$ WUs for $d = 3$.

Local Mode Analysis

6. **One-dimensional weighted Jacobi.**

    (a) Verify the Jacobi updating step (4.2).

    (b) Show that the error $e_j = u(x_j) - v_j$ satisfies (4.3).

    (c) Verify that the amplification factor for the method is given by

$$G(\theta) = 1 - 2\omega \sin^2\left(\frac{\theta}{2}\right).$$

7. **One-dimensional Gauss–Seidel.** Verify the error updating step (4.4). Then show that the amplification factor for the method is given by

$$G(\theta) = \frac{e^{i\theta}}{2 - e^{-i\theta}}.$$

8. **Two-dimensional weighted Jacobi.**

    (a) Verify the error updating step (4.6).

    (b) Show that the amplification factor for the method is given by

$$G(\theta_1, \theta_2) = 1 - \omega\left(\sin^2\left(\frac{\theta_1}{2}\right) + \sin^2\left(\frac{\theta_2}{2}\right)\right).$$

(c) Show that the optimal smoothing factor is obtained with $\omega = \frac{4}{5}$ and that its value is $\mu = |G(\pm\pi, \pm\pi)| = 0.6$. Hint: Note that $\mu(\omega) = \max |G(\theta_1, \theta_2)|$ is a function of ω. The optimal value of ω is that which minimizes μ, viewed as a function of ω. The substitutions $\xi = \sin^2(\frac{\theta_1}{2})$, $\eta = \sin^2(\frac{\theta_2}{2})$ may be helpful.

9. **Two-dimensional Gauss–Seidel.**

(a) Verify the error updating step (4.7).

(b) Show that the amplification factor for the method is given by

$$G(\theta_1, \theta_2) = \frac{e^{\iota\theta_1} + e^{\iota\theta_2}}{4 - e^{-\iota\theta_1} - e^{-\iota\theta_2}}.$$

(c) Show that the smoothing factor is given by

$$\mu = G\left(\frac{\pi}{2}, \cos^{-1}\left(\frac{4}{5}\right)\right) = \frac{1}{2}.$$

10. **Nine-point stencil.** Consider the nine-point stencil for the operator $-u_{xx} - u_{yy}$ given by

$$\frac{1}{3h^2}\begin{pmatrix} -1 & -1 & -1 \\ -1 & 8 & -1 \\ -1 & -1 & -1 \end{pmatrix}.$$

Find the amplification factors for the weighted Jacobi method and Gauss–Seidel relaxation applied to this system.

11. **Anisotropic operator.** Consider the five-point stencil for the operator $-\epsilon u_{xx} - u_{yy}$ given by

$$\frac{1}{h^2}\begin{pmatrix} 0 & -1 & 0 \\ -\epsilon & 2(1+\epsilon) & -\epsilon \\ 0 & -1 & 0 \end{pmatrix}.$$

Find the amplification factors for weighted Jacobi and Gauss–Seidel applied to this system. Discuss the effect of the parameter ϵ in the case that $\epsilon << 1$.

12. **Eigenvalue calculation in two dimensions.** Consider the weighted Jacobi method applied to the model Poisson equation in two dimensions on the unit square. Assume a uniform grid of $h = \frac{1}{n}$ in each direction.

(a) Let v_{ij} be the approximation to the solution at the grid point (x_i, y_j). Write the (i, j)th equation of the corresponding discrete problem, where $1 \leq i, j \leq n - 1$.

(b) Letting A be the matrix of coefficients for the discrete system, write the (i, j)th equation for the eigenvalue problem $A\mathbf{v} = \lambda\mathbf{v}$.

(c) Assume an eigenvector solution of the form

$$v_{ij} = \sin\left(\frac{ik\pi}{n}\right)\sin\left(\frac{j\ell\pi}{n}\right), \quad 1 \leq k, \ell \leq n - 1.$$

Using sine addition rules, simplify this eigenvalue equation, cancel common terms, and show that the eigenvalues are

$$\lambda_{k\ell} = 4\left[\sin^2\left(\frac{k\pi}{2n}\right) + \sin^2\left(\frac{\ell\pi}{2n}\right)\right], \quad 1 \leq k, \ell \leq n - 1.$$

(d) As in the one-dimensional case, note that the iteration matrix of the weighted Jacobi method is given by $P_\omega = I - \omega D^{-1} A$, where D corresponds to the diagonal terms of A. Find the eigenvalues of P_ω.

(e) Using a graphing utility, find a suitable way to present the two-dimensional set of eigenvalues (either a surface plot or multiple curves). Plot the eigenvalues for $\omega = \frac{2}{3}, \frac{4}{5}, 1$, and $n = 16$.

(f) In each case, discuss the effect of the weighted Jacobi method on low- and high-frequency modes. Be sure to note that modes can have a high frequencies in one direction and low frequencies in the other direction.

(g) What do you conclude about the optimal value of ω for the two-dimensional problem?

Implementation

13. **V-Cycle program.** Develop a V-cycle program for the one-dimensional model problem. Write a subroutine for each individual component of the algorithm as follows.

   (a) Given an approximation array **v**, a right-side array **f**, and a level number $1 \le l \le L$, write a subroutine that will carry out ν weighted Jacobi sweeps on level l.

   (b) Given an array **f** and a level number $1 \le l \le L - 1$, write a subroutine that will carry out full weighting between level l and level $l + 1$.

   (c) Given an array **v** and a level number $2 \le l \le L$, write a subroutine that will carry out linear interpolation between level l and level $l - 1$.

   (d) Write a driver program that initializes the data arrays and carries out a V-cycle by calling the three preceding subroutines. The program should be tested on simple problems for which the exact solution is known. For example, for fixed k, take $f(x) = C \sin(k\pi x)$ on the interval $0 \le x \le 1$, where C is a constant. Then

   $$u(x) = \frac{C}{\pi^2 k^2 + \sigma} \sin(k\pi x)$$

   is an exact solution to model problem (1.1). Another subroutine that computes norms of errors and residuals will be useful.

14. **Modification of V-cycle code.** It is now easy to modify this program and make comparisons.

   (a) Vary ν and vary the number of V-cycles.

   (b) Replace the weighted Jacobi subroutine, first by a Gauss–Seidel subroutine, then by a red-black Gauss–Seidel subroutine.

   (c) Replace the full weighting subroutine by an injection subroutine. Observe that using red-black Gauss–Seidel and injection impairs convergence. Try to remedy this by using half-injection. In all cases, determine experimentally how these changes affect convergence rates and computation time.

   (d) Explain the result of using black-red (rather than red-black) Gauss–Seidel with injection.

15. **Two-dimensional program.** For the two-dimensional problem, proceed again in a modular way.

   (a) Write a subroutine that performs weighted Jacobi on a two-dimensional grid. Within the subroutine, it is easiest to refer to **v** and **f** as two-dimensional arrays.

   (b) Make the appropriate modifications to the one-dimensional code to implement bilinear interpolation and full weighting on a two-dimensional grid.

   (c) Make the (minor) changes required in the main program to create a two-dimensional V-cycle program. Test this program on problems with known exact solutions. For example, for fixed k and ℓ, take $f(x,y) = C \sin(k\pi x) \sin(\ell\pi y)$ on the unit square $(0 \le x, y \le 1)$, where C is a constant. Then

   $$u(x,y) = \frac{C}{\pi^2 k^2 + \pi^2 \ell^2 + \sigma} \sin(k\pi x) \sin(\ell\pi y)$$

   is an exact soultion to model problem (1.4).

16. **FMG programs.** Modify the one- and two-dimensional V-cycle programs to carry out the FMG scheme.

17. **A convection-diffusion problem.** Consider the following convection-diffusion problem on the unit square $\Omega = \{(x,y) : 0 < x < 1, 0 < y < 1\}$:

   $$-\epsilon(u_{xx} + u_{yy}) + au_x = A\sin(\ell\pi y)(C_2 x^2 + C_1 x + C_0) \quad \text{on } \Omega,$$
   $$u = 0 \quad \text{on } \partial\Omega,$$

   where $\epsilon > 0$, $A \in \mathbf{R}$, $a \in \mathbf{R}$, ℓ is an integer, $C_2 = -\epsilon\ell^2\pi^2$, $C_1 = \epsilon\ell^2\pi^2 - 2a$, and $C_0 = a + 2\epsilon$. It has the exact solution $u(x,y) = Ax(1-x)\sin(\ell\pi y)$. Apply the multigrid algorithms discussed in this chapter to this problem. Compare the algorithms and explore how their performance changes for $\epsilon = 0.01, 0.1, 1$; $a = 0.1, 1, 10$; $n = 32, 64, 128$; $l = 1, 3, 16$.

18. **Discrete L^2 norm.** Let $u(x) = x^{m/2}$, where $m > -1$ is an integer, on $\Omega = [0, 1]$, with grid spacing $h = \frac{1}{n}$. Let $u_i^h = x_i^{m/2} = (ih)^{m/2}$. Show that the continuous L^2 norm is

   $$\|u\|_2 = \left(\int_0^1 x^{m/2}\right)^{1/2} = \frac{1}{\sqrt{m+1}},$$

   while the corresponding discrete L^2 norm satisfies

   $$\|\mathbf{u}^h\|_h \overset{h\to 0}{=} \frac{1}{\sqrt{m+1}}.$$

Chapter 5

Some Theory

In the previous chapter, we considered some practical questions concerning the implementation, cost, and performance of multigrid methods. The arguments and experiments of that chapter offer good reason to believe that multigrid methods can be extremely effective. Now we must confront some questions on the theoretical side. The goal of this chapter is to present multigrid in a more formal setting and offer an explanation of why these methods work so well. In the first part of this chapter, we sketch the ideas that underlie the convergence theory of multigrid. In the second section, we present what might be called the subspace picture of multigrid. While the terrain in this chapter may seem a bit more rugged than in previous chapters, the reward is an understanding of why multigrid methods are so remarkably effective.

Variational Properties

We first return to a question left unanswered in previous chapters. In expressing the coarse-grid problem, we wrote $A^{2h}u^{2h} = f^{2h}$ and said that A^{2h} is the Ω^{2h} version of the original operator A^h. Our first goal is to define the coarse-grid operator A^{2h} precisely.

The argument that follows assumes we are working with the model problem, $-u''(x) = f(x)$, and the corresponding discrete operator A^h. We adopt the notation that Ω^{ph} represents not only the grid with grid spacing ph, but also the space of vectors defined on that grid. In keeping with our former notation, we let v^h be a computed approximation to the exact solution u^h. For the purpose of this argument, assume that the error in this approximation, $e^h = u^h - v^h$, lies entirely in the range of interpolation, which will be denoted $\mathcal{R}(I^h_{2h})$. This means that for some vector $u^{2h} \in \Omega^{2h}$, $e^h = I^h_{2h}u^{2h}$. Therefore, the residual equation on Ω^h may be written

$$A^h e^h = A^h I^h_{2h} u^{2h} = r^h. \tag{5.1}$$

In this equation, A^h acts on a vector that lies entirely in the range of interpolation. How does A^h act on $\mathcal{R}(I^h_{2h})$? Figure 5.1 gives the answer. An arbitrary vector $u^{2h} \in \Omega^{2h}$ is shown in Fig. 5.1(a); $I^h_{2h}u^{2h}$, which is the interpolation of u^{2h} up to Ω^h, is shown in Fig. 5.1(b); and the effect of A^h acting pointwise on $I^h_{2h}u^{2h}$ is shown in Fig. 5.1(c). We see that $A^h I^h_{2h}u^{2h}$ is zero at the odd grid points of

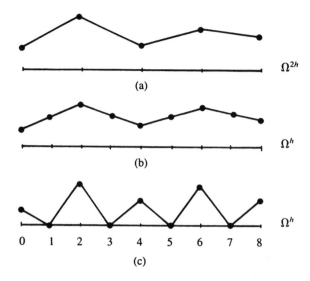

Figure 5.1: *The action of A^h on the range of interpolation $\mathcal{R}(I_{2h}^h)$: (a) an arbitrary vector $\mathbf{u}^{2h} \in \Omega^{2h}$; (b) the linear interpolant $I_{2h}^h \mathbf{u}^{2h}$; and (c) $A^h I_{2h}^h \mathbf{u}^{2h}$, which is zero at the odd grid points on Ω^h.*

Ω^h. The effect is analogous to taking the second derivative of a piecewise linear function.

We may conclude that the odd rows of $A^h I_{2h}^h$ in (5.1) are zero. On the other hand, the even rows of (5.1) correspond to the coarse-grid points of Ω^{2h}. Therefore, we can find a coarse-grid form of the residual equation by dropping the odd rows of (5.1). We do this formally by applying the restriction operator I_h^{2h} to both sides of (5.1). When this is done, the residual equation becomes

$$\underbrace{I_h^{2h} A^h I_{2h}^h}_{A^{2h}} \mathbf{u}^{2h} = I_h^{2h} \mathbf{r}^h.$$

This observation gives us a plausible definition for the coarse-grid operator: $A^{2h} = I_h^{2h} A^h I_{2h}^h$. The terms of A^{2h} may be computed explicitly as shown in Table 5.1. We simply apply $I_h^{2h} A^h I_{2h}^h$ term by term to the jth unit vector $\hat{\mathbf{e}}_j^{2h}$ on Ω^{2h}. This establishes that the jth column of A^{2h} and also, by symmetry, the jth row of A^{2h} are given by

$$\frac{1}{(2h)^2}\left(-1 \quad 2 \quad -1\right).$$

We would get the same result if the original problem were simply discretized on Ω^{2h} using the usual second-order finite differences. Therefore, by this definition, A^{2h} really is the Ω^{2h} version of A^h.

The preceding argument was based on the assumption that the error $\mathbf{e}^h$ lies entirely in the range of interpolation. This is not the case in general. If it were, then solving the Ω^{2h} residual equation exactly and doing the two-grid correction would give the exact solution. Nevertheless, the argument does give a sensible definition for A^{2h}. It also leads us to two important properties called the *variational*

	$j-1$		j		$j+1$
$\hat{\mathbf{e}}_j^{2h}$	0		1		0
$I_{2h}^h \hat{\mathbf{e}}_j^{2h}$	0	$\frac{1}{2}$	1	$\frac{1}{2}$	0
$A^h I_{2h}^h \hat{\mathbf{e}}_j^{2h}$	$-\frac{1}{2h^2}$	0	$\frac{1}{h^2}$	0	$-\frac{1}{2h^2}$
$I_h^{2h} A^h I_{2h}^h \hat{\mathbf{e}}_j^{2h}$	$-\frac{1}{4h^2}$		$\frac{1}{2h^2}$		$-\frac{1}{4h^2}$

Table 5.1: *Calculation of the ith row of $A^{2h} = I_h^{2h} A^h I_{2h}^h$.*

properties; they are given by

$$
\begin{aligned}
A^{2h} &= I_h^{2h} A^h I_{2h}^h \quad \text{(Galerkin condition)}, \\
I_h^{2h} &= c(I_{2h}^h)^T, \quad c \in \mathbf{R}.
\end{aligned}
$$

The first of these, the *Galerkin condition*, is simply the definition of the coarse-grid operator. The second property is the relationship satisfied by the interpolation operator and the full weighting operator defined in Chapter 3. While these properties are not desirable for all applications, they are exhibited by many commonly used operators. They also facilitate the analysis of the two-grid correction scheme. We see a deeper justification of these properties in Chapter 10.

Toward Convergence Theory

Convergence analysis of multigrid methods is difficult and has occupied researchers for several decades. We cannot even pretend to address this problem with the rigor and depth it deserves. Instead, we attempt to give heuristic arguments suggesting that the standard multigrid schemes, when applied to well-behaved problems (for example, scalar elliptic problems), not only work, but work very effectively. Convergence results for these problems have been rigorously proved. For more general problems, new results appear at a fairly steady pace. Where analytical results are lacking, a wealth of computational evidence testifies to the general effectiveness of multigrid methods. Between analysis and experimentation, the multigrid territory is slowly being mapped. However, multigrid convergence analysis is still an open area of computational mathematics.

We begin with a heuristic argument that captures the spirit of rigorous convergence proofs. As we have seen, the smoothing rate (the convergence factor for the oscillatory modes) for the standard relaxation schemes is small and independent of the grid spacing h. Recall that the smooth error modes, which remain after relaxation on one grid, appear more oscillatory on the coarser grids. Therefore, by moving to successively coarser grids, all of the error components on the original fine grid eventually appear oscillatory and are reduced by relaxation. It then follows that the overall convergence factor for a good multigrid scheme should be small and independent of h.

Now we can be a bit more precise. Denote the original continuous problem (for example, one of our model boundary value problems) by $Au = f$. The associated discrete problem on the fine grid Ω^h is denoted by $A^h \mathbf{u}^h = \mathbf{f}^h$. As before, we let $\mathbf{v}^h$ be an approximation to $\mathbf{u}^h$ on Ω^h. The *discretization error* is defined by

$$E_i^h = u(x_i) - u_i^h, \qquad 1 \le i \le n - 1.$$

The discretization error measures how well the exact solution of the discrete problem approximates the exact solution of the original continuous problem. It may be bounded in the discrete L^2 norm in the form

$$\|\mathbf{E}^h\|_h \le Kh^p, \tag{5.2}$$

where K is a positive constant and p is a positive integer. For the model problems in Chapter 1, in which second-order finite differences were used, we have $p = 2$ (see Exercise 11 for a careful derivation of this fact).

Unfortunately, we can seldom solve the discrete problem exactly. The quantity that we have been calling the error, $\mathbf{e}^h = \mathbf{u}^h - \mathbf{v}^h$, will now be called the *algebraic error* to avoid confusion with the discretization error. The algebraic error, as we have seen, measures how well our approximations (generated by relaxation or multigrid) agree with the exact discrete solution.

The purpose of a typical calculation is to produce approximations $\mathbf{v}^h$ that agree with the exact solution of the *continuous problem* $\mathbf{u}$. Let us specify a tolerance ϵ and an error condition such as

$$\|\mathbf{u} - \mathbf{v}^h\|_h < \epsilon, \tag{5.3}$$

where $\mathbf{u} = (u(x_1), \ldots, u(x_{n-1}))^T$ is the vector of exact solution values sampled at the grid points. This condition can be satisfied if we guarantee that both the discretization and algebraic errors are small. Suppose, for example, that

$$\|\mathbf{E}^h\|_h + \|\mathbf{e}^h\|_h < \epsilon.$$

Then, using the triangle inequality,

$$\|\mathbf{u} - \mathbf{v}^h\|_h \le \|\mathbf{u} - \mathbf{u}^h\|_h + \|\mathbf{u}^h - \mathbf{v}^h\|_h = \|\mathbf{E}^h\|_h + \|\mathbf{e}^h\|_h < \epsilon.$$

One way to ensure that $\|\mathbf{E}^h\|_h + \|\mathbf{e}^h\|_h < \epsilon$ is to require that $\|\mathbf{E}^h\|_h < \frac{\epsilon}{2}$ and $\|\mathbf{e}^h\|_h < \frac{\epsilon}{2}$ individually. The first condition determines the grid spacing on the finest grid. Using (5.2), it suggests that we choose

$$h < h^* \equiv \left(\frac{\epsilon}{2K}\right)^{1/p}.$$

The second condition determines how well $\mathbf{v}^h$ must approximate the exact discrete solution $\mathbf{u}^h$. If relaxation or multigrid cycles have been performed until the condition $\|\mathbf{e}^h\| < \frac{\epsilon}{2}$ is met on grid Ω^h, where $h < h^*$, then we have converged to the level of discretization error. In summary, the discretization error determines the critical grid spacing h^*; so (5.3) will be satisfied provided we converge to the level of discretization error on a grid with $h < h^*$.

Consider first a V-cycle scheme applied to a d-dimensional problem with (about) n^d unknowns and $h = \frac{1}{n}$. We assume (and can generally show rigorously) that with fixed cycling parameters, ν_1 and ν_2, the V-cycle scheme has a convergence factor bound, γ, that is independent of h. This V-cycle scheme must reduce the algebraic error from $O(1)$ (the error in the zero initial guess) to $O(h^p) = O(n^{-p})$ (the order of the discretization error). Therefore, the number of V-cycles required, ν, must satisfy $\gamma^\nu = O(n^p)$ or $\nu = O(\log n)$. Because the cost of a single V-cycle is $O(n^d)$, the cost of converging to the level of discretization error with a V-cycle scheme is $O(n^d \log n)$. This is comparable to the computational cost of the best fast direct solvers applied to the model problem.

The FMG scheme costs a little more per cycle than the V-cycle scheme. However, a properly designed FMG scheme can be much more effective overall because it supplies a very good initial guess to the final V-cycles on Ω^h. The key observation in the FMG argument is that before the Ω^h problem is even touched, the Ω^{2h} problem has already been solved to the level of discretization error. This is because of nested iteration, which is designed to provide a good initial guess for the next finer grid. Our goal is to determine how much the algebraic error needs to be reduced by V-cycles on Ω^h. The argument is brief and worthwhile; however, it requires a new tool.

Energy Norms, Inner Products, and Orthogonality. Energy norms and inner products are useful tools in the analysis of multigrid methods. They involve a slight modification of the inner product and norms that we have already encountered. Suppose A is an $n \times n$ symmetric positive definite matrix. Define the A-*inner product* of two vectors $\mathbf{u}, \mathbf{v} \in \mathbf{R}^n$ by

$$(\mathbf{u}, \mathbf{v})_A \equiv (A\mathbf{u}, \mathbf{v}),$$

where $(\cdot, \cdot)$ is the usual Euclidean inner product on $\mathbf{R}^n$. The A-*norm* now follows in a natural way. Just as $\|\mathbf{u}\| = (\mathbf{u}, \mathbf{u})^{\frac{1}{2}}$, the A-norm is given by

$$\|\mathbf{u}\|_A = (\mathbf{u}, \mathbf{u})_A^{\frac{1}{2}}.$$

The A-norm and A-inner product are sometimes called the *energy* norm and inner product. We can also use the A-inner product to define a new orthogonality relationship. Extending the usual notion of vectors and subspaces, two vectors u and v are A-*orthogonal* if $(u, v)_A = 0$, and we write $u \perp_A v$. We then say that two subspaces U and V are A-orthogonal if, for all $u \in U$ and $v \in V$, we have $u \perp_A v$. In this case, we write $U \perp_A V$.

Our FMG argument can be made in any norm, but it is simplest in the A^h-norm. The goal is to show that one properly designed FMG cycle is enough to ensure that the final algebraic error on Ω^h is below the level of discretization error; that is,

$$\|\mathbf{e}^h\|_{A^h} \le Kh^p, \tag{5.4}$$

where K is a positive constant that depends on the smoothness of the solution and p is a positive integer. The values of K and p also depend on the norm used to measure the error, so they will generally be different from the constants in (5.2).

The argument is inductive in nature. If Ω^h is the coarsest grid, then FMG is exact and (5.4) is clearly satisfied. Assume now that the Ω^{2h} problem has been solved to the level of discretization error, so that

$$\|\mathbf{e}^{2h}\|_{A^{2h}} \leq K(2h)^p. \tag{5.5}$$

We now use (5.5) to prove (5.4).

The initial algebraic error on Ω^h, before the V-cycles begin, is the difference between the exact fine-grid solution, $\mathbf{u}^h$, and the coarse-grid approximation interpolated to the fine grid:

$$\mathbf{e}_0^h = \mathbf{u}^h - I_{2h}^h \mathbf{v}^{2h}.$$

To estimate the size of this initial error, we must account for the error that might be introduced by interpolation from Ω^{2h} to Ω^h. To do this, we assume that the error in interpolation has the same order (same value of p) as the discretization error and satisfies

$$\|\mathbf{u}^h - I_{2h}^h \mathbf{u}^{2h}\|_{A^h} \leq K\alpha h^p, \tag{5.6}$$

where α is a positive constant. (This sort of bound can be determined rigorously; in fact, α is typically $1 + 2^p$.) A bound for the initial error now follows from the triangle inequality:

$$
\begin{aligned}
\|\mathbf{e}_0^h\|_{A^h} &= \|\mathbf{u}^h - I_{2h}^h \mathbf{v}^{2h}\|_{A^h} \\
&\leq \|\mathbf{u}^h - I_{2h}^h \mathbf{u}^{2h}\|_{A^h} + \|I_{2h}^h \mathbf{u}^{2h} - I_{2h}^h \mathbf{v}^{2h}\|_{A^h} \quad \text{(triangle inequality)} \\
&= \underbrace{\|\mathbf{u}^h - I_{2h}^h \mathbf{u}^{2h}\|_{A^h}}_{\leq K\alpha h^p} + \underbrace{\|\mathbf{u}^{2h} - \mathbf{v}^{2h}\|_{A^{2h}}}_{\leq K(2h)^p} \quad \text{(Galerkin conditions)}.
\end{aligned}
$$

As indicated, we use (5.5), (5.6), and Exercise 2 to form the following estimate for the norm of the initial error:

$$\|\mathbf{e}_0^h\|_{A^h} \leq K\alpha h^p + K(2h)^p = K(\alpha + 2^p)h^p.$$

To satisfy (5.4), the algebraic error must be reduced from roughly $K(\alpha + 2^p)h^p$ to Kh^p. This means we should use enough V-cycles on Ω^h to reduce the algebraic error by a factor of

$$\beta = \alpha + 2^p.$$

This reduction requires ν V-cycles, where $\gamma^\nu \leq \beta$. Because β is $O(1)$, it follows that ν is also $O(1)$. (Typically, $\beta = 5$ or 9 and $\gamma \approx 0.1$ for a V(2,1)-cycle, so $\nu = 1$.) In other words, the preliminary cycling through coarser grids gives such a good initial guess that only $O(1)$ V-cycles are needed on the finest grid. This means that the total computational cost of FMG is $O(n^d)$, which is optimal.

This discussion is meant to give some feeling for the rigorous arguments that can be used to establish the convergence properties of the basic multigrid algorithms. These results cannot be pursued much further at this point and must be left to the multigrid literature. Instead, we turn to a different perspective on why multigrid works.

Spectral and Algebraic Pictures

Much of this section is devoted to an analysis of the two-grid correction scheme, which lies at the heart of multigrid. Recall that the V-cycle is just nested applications of the two-grid correction scheme and that the FMG method is just

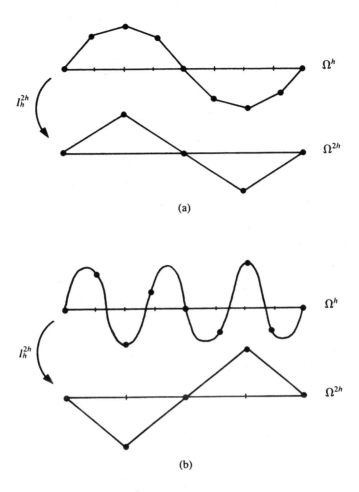

Figure 5.2: *The full weighting operator I_h^{2h} acting on* (a) *a smooth mode of Ω^h* *($k = 2$ and $n = 8$) and* (b) *an oscillatory mode of Ω^h ($k = 6$ and $n = 8$). In the first case, the result is a multiple of the $k = 2$ mode on Ω^{2h}. In the second case, the result is a multiple of the $n - k = 2$ mode on Ω^{2h}.*

repeated applications of the V-cycle on various grids. Therefore, an understanding of the two-grid correction scheme is essential for a complete explanation of the basic multigrid methods.

We begin with a detailed look at the intergrid transfer operators. Consider first the full weighting (restriction) operator, I_h^{2h}. Recall that I_h^{2h} maps $\mathbf{R}^{n-1} \to \mathbf{R}^{\frac{n}{2}-1}$. It has rank $\frac{n}{2} - 1$ and null space $N(I_h^{2h})$ of dimension $\frac{n}{2}$. It is important to understand what we call the spectral properties of I_h^{2h}. In particular, how does I_h^{2h} act upon the modes of the original operator A^h?

Recall that the modes of A^h for the one-dimensional model problem are given by

$$w_{k,j}^h = \sin\left(\frac{jk\pi}{n}\right), \qquad 1 \le k \le n - 1, \quad 0 \le j \le n.$$

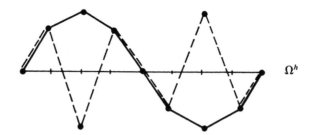

Ω^h

Figure 5.3: *A pair of complementary modes on a grid with $n = 8$ points. The solid line shows the $k = 2$ mode. The dashed line shows the $k' = n - k = 6$ mode.*

The full weighting operator may be applied directly to these vectors. The result of I_h^{2h} acting on the smooth modes is (Exercise 4)

$$I_h^{2h} \mathbf{w}_k^h = \cos^2 \left(\frac{k\pi}{2n} \right) \mathbf{w}_k^{2h}, \qquad 1 \le k \le \frac{n}{2}.$$

This says that I_h^{2h} acting on the kth (smooth) mode of A^h produces a constant times the kth mode of A^{2h} when $1 \le k \le \frac{n}{2}$. This property is illustrated in Fig. 5.2(a). For the oscillatory modes, with $\frac{n}{2} < k < n - 1$, we have (Exercise 5)

$$I_h^{2h} \mathbf{w}_{k'}^h = -\sin^2 \left(\frac{k\pi}{2n} \right) \mathbf{w}_k^{2h}, \qquad 1 \le k < \frac{n}{2},$$

where $k' = n - k$. This says that I_h^{2h} acting on the $(n - k)$th mode of A^h produces a constant multiple of the kth mode of A^{2h}. This property, illustrated in Fig. 5.2(b), is similar to the aliasing phenomenon discussed earlier. The oscillatory modes on Ω^h cannot be represented on Ω^{2h}. As a result, the full weighting operator transforms these modes into relatively smooth modes on Ω^{2h}.

In summary, we see that both the kth and $(n - k)$th modes on Ω^h become the kth mode on Ω^{2h} under the action of full weighting. We refer to this pair of fine-grid modes $\{\mathbf{w}_k^h, \mathbf{w}_{n-k}^h\}$ as *complementary modes*. Letting $W_k^h = \text{span}\{\mathbf{w}_k^h, \mathbf{w}_{n-k}^h\}$, we have that

$$I_h^{2h} : W_k^h \to \text{span}\{\mathbf{w}_k^{2h}\}.$$

In passing, it is interesting to note the relationship between two complementary modes. It may be shown (Exercise 6) that $w_{n-k,j}^h = (-1)^{j+1} w_{k,j}^h$. Figure 5.3 illustrates the smooth and oscillatory nature of a pair of complementary modes.

As noted earlier, the full weighting operator has a nontrivial null space, $N(I_h^{2h})$. It may be verified (Exercise 7) that this subspace is spanned by the vectors $\mathbf{n}_j = A^h \hat{\mathbf{e}}_j^h$, where j is odd and $\hat{\mathbf{e}}_j^h$ is the jth unit vector on Ω^h. As shown in Fig. 5.4, the basis vectors $\mathbf{n}_j$ appear oscillatory. However, they do not coincide with the oscillatory modes of A^h. In fact, an expansion of $\mathbf{n}_j$ in terms of the modes of A^h requires all of the modes. Thus, the null space of I_h^{2h} possesses both smooth and oscillatory modes of A^h.

Having established the necessary properties of the full weighting operator, we now examine the interpolation operator I_{2h}^h in the same way. Recall that I_{2h}^h maps

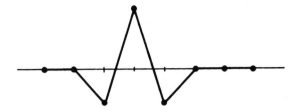

Figure 5.4: *A typical basis vector of the null space of the full weighting operator* $N(I_h^{2h})$.

Figure 5.5: *A typical basis vector of the range of interpolation* $\mathcal{R}(I_{2h}^h)$.

$\mathbf{R}^{\frac{n}{2}-1} \to \mathbf{R}^{n-1}$ and has full rank. In order to establish the spectral properties of I_{2h}^h, we ask how I_{2h}^h acts on the modes of A^{2h}. Letting

$$w_{k,j}^{2h} = \sin\left(\frac{jk\pi}{n/2}\right), \qquad 1 \le k < \frac{n}{2}, \quad 0 \le j \le \frac{n}{2},$$

be the Ω^{2h} modes, we can show (Exercise 8) that I_{2h}^h does not preserve these modes. The calculation shows that

$$I_{2h}^h \mathbf{w}_k^{2h} = c_k \mathbf{w}_k^h - s_k \mathbf{w}_{k'}^h, \qquad 1 \le k < \frac{n}{2}, \quad k' = n - k,$$

where $c_k = \cos^2\left(\frac{k\pi}{2n}\right)$ and $s_k = \sin^2\left(\frac{k\pi}{2n}\right)$. We see that I_{2h}^h acting on the kth mode of Ω^{2h} produces not only the kth mode of Ω^h but also the complementary mode $\mathbf{w}_{k'}^h$. This fact exposes the interesting property that interpolation of smooth modes on Ω^{2h} excites (to some degree) oscillatory modes on Ω^h. It should be noted that for a very smooth mode on Ω^h with $k \ll n/2$,

$$I_{2h}^h \mathbf{w}_k^{2h} = \left[1 - O\left(\frac{k^2}{n^2}\right)\right] \mathbf{w}_k^h + O\left(\frac{k^2}{n^2}\right) \mathbf{w}_{k'}^h.$$

In this case, the result of interpolation is largely the corresponding smooth mode on Ω^h with very little contamination from the complementary oscillatory mode. As it has been defined, I_{2h}^h is a *second-order* interpolation operator because the magnitude of the spurious oscillatory mode is $O(\frac{k^2}{n^2})$.

We have already anticipated the importance of the range of interpolation, $\mathcal{R}(I_{2h}^h)$. A basis for $\mathcal{R}(I_{2h}^h)$ is given by the columns of I_{2h}^h. While these basis vectors appear smooth, as Fig. 5.5 shows, they do not coincide with the smooth modes of A^h. In fact, it may be shown that any one of these basis vectors requires all modes of A^h for a full representation. In other words, the range of interpolation contains both smooth and oscillatory modes of A^h.

With this investigation of the intergrid transfer operators, we now return to the two-grid correction scheme. We begin with an observation made in Chapter 2 that a stationary linear iteration may be expressed in the form

$$\mathbf{v}^{(1)} = (I - BA)\mathbf{v}^{(0)} + B\mathbf{f} = R\mathbf{v}^{(0)} + B\mathbf{f},$$

where B is a specified matrix and $R = I - BA$ is the iteration matrix for the method. It follows that m sweeps of the iteration can be represented by

$$\mathbf{v}^{(m)} = R^m \mathbf{v}^{(0)} + C(\mathbf{f}),$$

where $C(\mathbf{f})$ represents a series of operations on $\mathbf{f}$.

We can now turn to the two-grid correction scheme. The steps of this scheme, with an exact solution on the coarse grid, are given by the following procedure:

- Relax ν times on Ω^h with scheme R: $\mathbf{v}^h \leftarrow R^\nu \mathbf{v}^h + C(\mathbf{f})$.

- Full weight $\mathbf{r}^h$ to Ω^{2h}: $\mathbf{f}^{2h} \leftarrow I_h^{2h}(\mathbf{f}^h - A^h \mathbf{v}^h)$.

- Solve the residual equation exactly: $\mathbf{v}^{2h} = (A^{2h})^{-1}\mathbf{f}^{2h}$.

- Correct the approximation on Ω^h: $\mathbf{v}^h \leftarrow \mathbf{v}^h + I_{2h}^h \mathbf{v}^{2h}$.

If we now take this process one step at a time, it may be represented in terms of a single replacement operation:

$$\mathbf{v}^h \leftarrow R^\nu \mathbf{v}^h + C(\mathbf{f}) + I_{2h}^h (A^{2h})^{-1} I_h^{2h}(\mathbf{f}^h - A^h(R^\nu \mathbf{v}^h + C(\mathbf{f}))).$$

The exact solution $\mathbf{u}^h$ is unchanged by the two-grid correction scheme. Therefore,

$$\mathbf{u}^h = R^\nu \mathbf{u}^h + C(\mathbf{f}) + I_{2h}^h (A^{2h})^{-1} I_h^{2h}(\mathbf{f}^h - A^h(R^\nu \mathbf{u}^h + C(\mathbf{f}))).$$

By subtracting these last two expressions, we can see how the two-grid correction operator, which we now denote TG, acts upon the error, $\mathbf{e}^h = \mathbf{u}^h - \mathbf{v}^h$. We find that

$$\mathbf{e}^h \leftarrow \left[I - I_{2h}^h (A^{2h})^{-1} I_h^{2h} A^h \right] R^\nu \mathbf{e}^h \equiv TG \mathbf{e}^h. \tag{5.7}$$

As in Chapter 2, we imagine that the error can be expressed as a linear combination of the modes of A^h. This leads us to ask how TG acts upon the modes of A^h. However, TG consists of R, A^h, $(A^{2h})^{-1}$, I_h^{2h}, and I_{2h}^h, and we now know how each of these operators acts upon the modes of A^h. For the moment, consider the two-grid correction scheme TG with no relaxation ($\nu = 0$). Using all of the spectral properties we have just discovered, it may be shown (Exercise 9) that the coarse-grid correction operator, TG, is invariant on the subspaces $W_k^h = \text{span}\{\mathbf{w}_k^h, \mathbf{w}_{k'}^h\}$; that is,

$$TG\mathbf{w}_k = s_k \mathbf{w}_k + s_k \mathbf{w}_{k'}, \tag{5.8}$$

$$TG\mathbf{w}_{k'} = c_k \mathbf{w}_k + c_k \mathbf{w}_{k'}, \quad 1 \le k \le \frac{n}{2}, \quad k' = n - k, \tag{5.9}$$

where $c_k = \cos^2\left(\frac{\pi k}{2n}\right)$ and $s_k = \sin^2\left(\frac{\pi k}{2n}\right)$.

This implies that when TG is applied to a smooth or oscillatory mode, the same mode and its complement result. But it is important to look at the amplitudes of

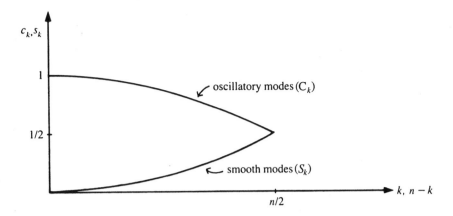

Figure 5.6: *Damping factor for the two-grid correction operator TG, without relaxation, acting on the oscillatory modes with wavenumbers $n - k$ (upper curve) and on the smooth modes with wavenumbers k (lower curve) for $1 \le k \le \frac{n}{2}$.*

the resulting modes. Consider the case of TG acting on very smooth modes and very oscillatory modes with $k \ll n$. Then (5.8) and (5.9) become

$$TG\mathbf{w}_k = O\left(\frac{k^2}{n^2}\right)\mathbf{w}_k + O\left(\frac{k^2}{n^2}\right)\mathbf{w}_{k'},$$

$$TG\mathbf{w}_{k'} = \left[1 - O\left(\frac{k^2}{n^2}\right)\right]\mathbf{w}_k + \left[1 - O\left(\frac{k^2}{n^2}\right)\right]\mathbf{w}_{k'}, \quad 1 \le k \le \frac{n}{2}, \quad k' = n - k.$$

TG acting on smooth modes produces smooth and oscillatory modes with very small amplitudes. Therefore, the two-grid correction scheme is effective at eliminating smooth components of the error. However, when TG acts upon highly oscillatory modes, it produces smooth and oscillatory modes with $O(1)$ amplitudes. Therefore, two-grid correction, without relaxation, does not damp oscillatory modes. Figure 5.6 illustrates this behavior of the two-grid correction scheme with no relaxation by showing the damping factors, c_k and s_k, for the smooth and oscillatory components of the error.

We now bring relaxation into the picture. Knowing its spectral properties, we can anticipate that relaxation will balance perfectly the action of TG without relaxation. We now include ν steps of a relaxation method R and assume for simplicity that R does not mix the modes of A^h. Many other relaxation methods can be analyzed without this assumption. As before, let λ_k be the eigenvalue of R associated with the kth mode $\mathbf{w}_k$. Combining all of these observations with (5.7), the action of TG *with* relaxation is given by (Exercise 10)

$$TG\mathbf{w}_k = \lambda_k^\nu s_k \mathbf{w}_k + \lambda_k^\nu s_k \mathbf{w}_{k'}, \tag{5.10}$$

$$TG\mathbf{w}_{k'} = \lambda_{k'}^\nu c_k \mathbf{w}_k + \lambda_{k'}^\nu c_k \mathbf{w}_{k'}, \quad 1 \le k \le \frac{n}{2}, \; k' = n - k. \tag{5.11}$$

We know that the smoothing property of relaxation has the strongest effect on the oscillatory modes. This is reflected in the term $\lambda_{k'}^\nu$, which is small. At the same time, the two-grid correction scheme alone (without relaxation) eliminates the smooth modes. This is reflected in the s_k terms. Thus, all terms of (5.10) and (5.11) are small, particularly for $k \ll \frac{n}{2}$ or as ν becomes large. The result is a complete process in that both smooth and oscillatory modes of the error are well damped.

We have now completed what we call the spectral picture of multigrid. By examining how various operators act on the modes of A^h, we have determined the effect of the entire two-grid correction operator on those modes. This analysis explains how the two-grid correction process eliminates both the smooth and oscillatory components of the error.

Solvability and the Fundamental Theorem of Linear Algebra. Suppose we have a matrix $A \in \mathbf{R}^{m \times n}$. The fundamental theorem of linear algebra states that the range (column space) of the matrix, $\mathcal{R}(A)$, is equal to the orthogonal complement of $\mathcal{N}(A^T)$, the null space of A^T. Thus, spaces $\mathbf{R}^m$ and $\mathbf{R}^n$ can be orthogonally decomposed as follows:

$$\begin{aligned} \mathbf{R}^m &= \mathcal{R}(A) \oplus \mathcal{N}(A^T), \\ \mathbf{R}^n &= \mathcal{R}(A^T) \oplus \mathcal{N}(A). \end{aligned}$$

For the equation $A\mathbf{x} = \mathbf{f}$ to have a solution, it is necessary that the vector $\mathbf{f}$ lie in $\mathcal{R}(A)$. Thus, an equivalent condition is that $\mathbf{f}$ be orthogonal to every vector in $\mathcal{N}(A^T)$. For the equation $A\mathbf{x} = \mathbf{f}$ to have a *unique* solution, it is necessary that $\mathcal{N}(A) = \{\mathbf{0}\}$. Otherwise, if $\mathbf{x}$ is a solution and $\mathbf{y} \in \mathcal{N}(A)$, then $A(\mathbf{x} + \mathbf{y}) = A\mathbf{x} + A\mathbf{y} = \mathbf{f} + \mathbf{0} = \mathbf{f}$, so the solution $\mathbf{x}$ is not unique.

There is another vantage point from which to view the coarse-grid correction scheme. This perspective will lead to what we will call the algebraic picture of multigrid. With both the spectral and the algebraic picture before us, it will be possible to give a good qualitative explanation of multigrid. Let us now look at the algebraic structure of the two-grid correction scheme.

The variational properties introduced earlier now become important. Recall that these properties are given by

$$\begin{aligned} A^{2h} &= I_h^{2h} A^h I_{2h}^h \qquad \text{(Galerkin property)}, \\ I_h^{2h} &= c(I_{2h}^h)^T, \qquad c \in \mathbf{R}. \end{aligned}$$

The two-grid correction scheme involves transformations between the space of fine-grid vectors, Ω^h, and the space of coarse-grid vectors, Ω^{2h}. Figure 5.7 diagrams these two spaces and the action of the full weighting and interpolation operators.

As we have already seen, the range of interpolation, $\mathcal{R}(I_{2h}^h)$, and the null space of full weighting, $N(I_h^{2h})$, both reside in Ω^h and have dimensions of roughly $\frac{n}{2}$. From the orthogonality relationships between the subspaces of a linear operator (Fundamental Theorem of Linear Algebra), we know that

$$N(I_h^{2h}) \perp \mathcal{R}[(I_h^{2h})^T].$$

By the second variational property, it then follows that

$$N(I_h^{2h}) \perp \mathcal{R}(I_{2h}^h).$$

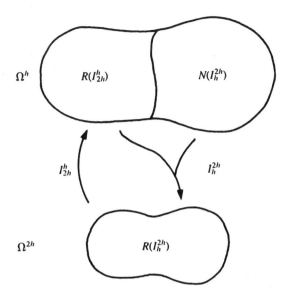

Figure 5.7: *Schematic drawing of the space of fine-grid vectors, Ω^h; the space of coarse-grid vectors, Ω^{2h}; and their subspaces and the intergrid transfer operators, I^h_{2h} and I^{2h}_h.*

The fact that the range of interpolation is orthogonal to the null space of full weighting is significant.

We will now use the notion of A-orthogonality to rewrite the above orthogonality relationship. The fact that $N(I^{2h}_h) \perp \mathcal{R}(I^h_{2h})$ means that $(\mathbf{q}^h, \mathbf{u}^h) = 0$ whenever $\mathbf{q}^h \in \mathcal{R}(I^h_{2h})$ and $I^{2h}_h\mathbf{u}^h = \mathbf{0}$. This is equivalent to the condition that $(\mathbf{q}^h, A^h\mathbf{u}^h) = 0$ whenever $\mathbf{q}^h \in \mathcal{R}(I^h_{2h})$ and $I^{2h}_h A^h\mathbf{u}^h = \mathbf{0}$. This last condition may be written as

$$N(I^{2h}_h A^h) \perp_{A^h} \mathcal{R}(I^h_{2h});$$

that is, the null space of $I^{2h}_h A^h$ is A^h-orthogonal to the range of interpolation.

This orthogonality property allows the space Ω^h to be decomposed in the form

$$\Omega^h = \mathcal{R}(I^h_{2h}) \oplus N(I^{2h}_h A^h).$$

This means that if $\mathbf{e}^h$ is a vector in Ω^h, then it may always be expressed as

$$\mathbf{e}^h = \mathbf{s}^h + \mathbf{t}^h,$$

where $\mathbf{s}^h \in \mathcal{R}(I^h_{2h})$ and $\mathbf{t}^h \in N(I^{2h}_h A^h)$.

It will be helpful to interpret the vectors $\mathbf{s}^h$ and $\mathbf{t}^h$. Since $\mathbf{s}^h$ is an element of $\mathcal{R}(I^h_{2h})$, it must satisfy $\mathbf{s}^h = I^h_{2h}\mathbf{q}^{2h}$, where $\mathbf{q}^{2h}$ is some vector of Ω^{2h}. We observed the smoothing effect of interpolation and noted the smooth appearance of the basis vectors of $\mathcal{R}(I^h_{2h})$. For this reason, we associate $\mathbf{s}^h$ with the smooth components of the error. We also noted the oscillatory appearance of the basis vectors of $N(I^{2h}_h)$. For this reason, we associate $\mathbf{t}^h$ with the oscillatory components of the error.

We may now consider the two-grid correction operator in light of these subspace properties. The two-grid correction operator without relaxation is

$$TG = I - I^h_{2h}(A^{2h})^{-1}I^{2h}_h A^h.$$

First note that if $\mathbf{s}^h \in \mathcal{R}(I_{2h}^h)$, then $\mathbf{s}^h = I_{2h}^h \mathbf{q}^{2h}$ for some vector $\mathbf{q}^{2h}$ in Ω^{2h}. We then have that

$$TG\mathbf{s}^h = \left[I - I_{2h}^h (A^{2h})^{-1} I_h^{2h} A^h\right] I_{2h}^h \mathbf{q}^{2h}. \tag{5.12}$$

However, by the first of the variational properties, $I_h^{2h} A^h I_{2h}^h = A^{2h}$. Therefore, $TG\mathbf{s}^h = \mathbf{0}$. This gives us the important result that any vector in the range of interpolation also lies in the null space of the two-grid correction operator, that is,

$$N(TG) \supset \mathcal{R}(I_{2h}^h).$$

Having seen how TG acts on $\mathcal{R}(I_{2h}^h)$, now consider a vector $\mathbf{t}^h$ in $N(I_h^{2h} A^h)$. This case is even simpler. We have

$$TG\mathbf{t}^h = \left[I - I_{2h}^h (A^{2h})^{-1} I_h^{2h} A^h\right] \mathbf{t}^h. \tag{5.13}$$

Because $I_h^{2h} A^h \mathbf{t}^h = \mathbf{0}$, we conclude that $TG\mathbf{t}^h = \mathbf{t}^h$. This says that TG is the identity when it acts on $N(I_h^{2h} A^h)$. This implies that the dimension of $N(TG)$ cannot exceed the dimension of $\mathcal{R}(I_{2h}^h)$. We therefore have the stronger result (Exercise 11) that

$$N(TG) = \mathcal{R}(I_{2h}^h).$$

This argument gives us the effect of the two-grid correction operator on the two orthogonal subspaces of Ω^h. Let us now put these algebraic results together with the spectral picture. We have established two independent ways to decompose the space of fine-grid vectors, Ω^h. We have the spectral decomposition

$$\Omega^h = L \oplus H = \left\{ \begin{array}{c} \text{Low-frequency modes} \\ 1 \leq k < \dfrac{n}{2} \end{array} \right\} \oplus \left\{ \begin{array}{c} \text{High-frequency modes} \\ \dfrac{n}{2} \leq k < n \end{array} \right\}$$

and the subspace decomposition

$$\Omega^h = \mathcal{R}(I_{2h}^h) \oplus N(I_h^{2h} A^h).$$

We can now give a schematic illustration of the two-grid correction scheme as it works on an arbitrary error vector. The diagrams in Figs. 5.8–5.11 are rather unconventional and may require some deliberation. However, they do incorporate all of the spectral and subspace results elaborated on in this chapter.

We first focus on the upper diagrams of each figure. These diagrams portray the space Ω^h as the plane of the page. As we have seen, Ω^h may be decomposed in two ways. These two decompositions are represented by two pairs of orthogonal axes labeled (H, L) for the high-frequency/low-frequency decomposition and (N, R) for the $N(I_h^{2h} A^h)/\mathcal{R}(I_{2h}^h)$ decomposition. The L and R axes are more nearly aligned, suggesting that the smooth, low-frequency modes are associated with the range of interpolation.

Initially, an arbitrary error vector $\mathbf{e}^h$ in Ω^h appears as a point in the plane as shown in Fig. 5.8. This vector has projections on all four axes. Specifically, consider the projections on the R and N axes, which we have called $\mathbf{s}^h$ and $\mathbf{t}^h$, respectively. Both $\mathbf{s}^h$ and $\mathbf{t}^h$ may be further projected onto the L and H axes. The projections of $\mathbf{s}^h$ on the L and H axes are denoted $\mathbf{s}_L$ and $\mathbf{s}_H$, respectively. The projections of $\mathbf{t}^h$ on the L and H axes are denoted $\mathbf{t}_L$ and $\mathbf{t}_H$, respectively. This gives the picture for the initial error, before any relaxation is done.

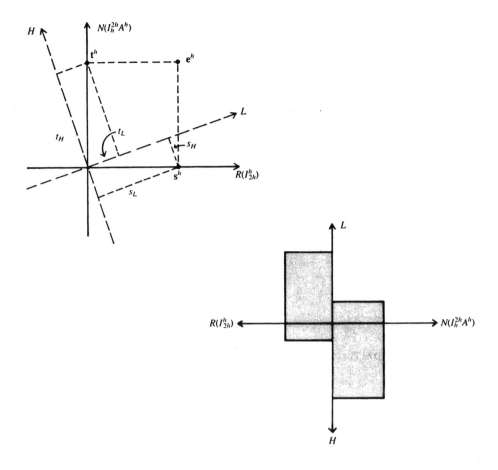

Figure 5.8: *The top diagram shows the space Ω^h decomposed along the (low-frequency, high-frequency) axes and along the $(\mathcal{R}(I_{2h}^h), N(I_h^{2h}A^h))$ axes. The initial error, $\mathbf{e}^h$, appears as a point with projections on all four axes. The lower diagram gives a schematic "energy budget" of the initial error. The error energy is divided between $\mathcal{R}(I_{2h}^h)$ and $N(I_h^{2h}A^h)$ and further divided between low- and high-frequency modes.*

The upper diagram in Fig. 5.9 shows the effect of several relaxation sweeps on the fine grid. Assume that enough relaxation sweeps are done to eliminate entirely the high-frequency components of the error. In the diagram, the resulting error has no component along the H-axis and the point representing $\mathbf{e}^h$ is projected down to the L-axis. The diagram also shows the new projections on the R and N axes. Notice that the component of $\mathbf{e}^h$ in $\mathcal{R}(I_{2h}^h)$ has actually increased.

The upper diagram of Fig. 5.10 shows the effect of the rest of the two-grid correction scheme. Since $\mathcal{R}(I_{2h}^h)$ is the null space of TG, the component of $\mathbf{e}^h$ along the R-axis vanishes. Therefore, this step is represented by a projection directly onto the N-axis. This is consistent with our observation that TG acts as the identity on $N(I_h^{2h}A^h)$. This new projection has eliminated a large amount of the initial error. However, as we proved earlier, the coarse-grid correction operator (without

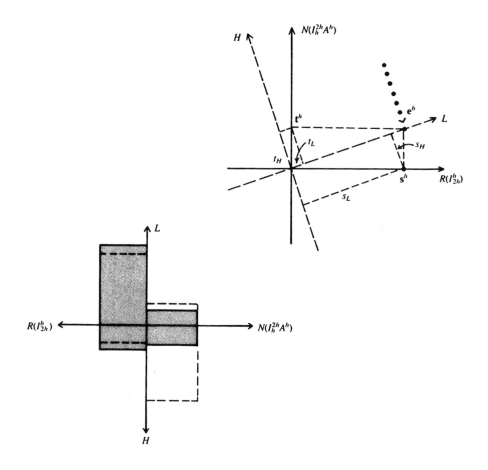

Figure 5.9: *The top diagram shows the effect of several fine-grid relaxation sweeps. Assuming that all oscillatory components are eliminated, the error is projected onto the low-frequency axis (L). The lower diagram shows the corresponding changes in the energy budget. Dashed lines indicate the previous energy budget. Shaded areas show the current budget.*

relaxation) does excite oscillatory modes, which can be seen in the diagram as a nonzero component in the H direction.

The upper diagram of Fig. 5.11 shows the error after more relaxation sweeps on the fine grid. Assuming again that relaxation eliminates all high-frequency error modes, this relaxation sweep is represented as a projection directly onto the L-axis. This further reduces a significant amount of the error.

The pattern should now be evident. By combining relaxation with corrections from the coarse-grid residual equation, we alternately project onto the L and N axes. In doing this, the error vector is driven toward the origin (zero error) in a way that is reminiscent of the convergence of a fixed point iteration.

The lower diagrams of Figs. 5.8–5.11 attempt to illustrate the two-grid correction scheme in terms of the "energy budget" of the error. Initially, the error has a certain energy or magnitude. Part of this energy resides in $\mathcal{R}(I_{2h}^h)$, which is drawn

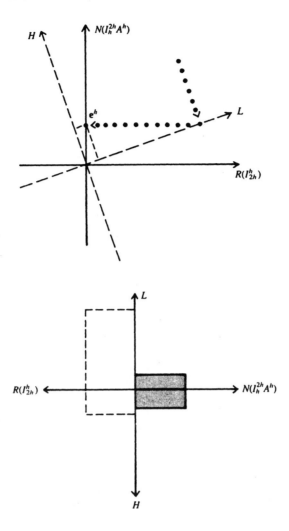

Figure 5.10: *The top diagram shows the error after doing the remainder of the two-grid correction. All components in $\mathcal{R}(I_{2h}^h)$ are eliminated as indicated by the projection onto the N-axis. The lower diagram shows the corresponding changes in the energy budget.*

on the left of the vertical line; the remainder of the energy resides in $N(I_h^{2h}A^h)$, which is drawn on the right of that line. The energy in $\mathcal{R}(I_{2h}^h)$ may be divided further between the low-frequency modes (above the horizontal line) and the high-frequency modes (below the horizontal line). We see that the initial error is fairly evenly divided between $\mathcal{R}(I_{2h}^h)$ and $N(I_h^{2h}A^h)$. As we noted before, most of the error in $\mathcal{R}(I_{2h}^h)$ is associated with the smooth, low-frequency modes; most of the error in $N(I_h^{2h}A^h)$ is associated with the oscillatory, high-frequency modes.

The lower diagram of Fig. 5.9 shows the energy budget after several effective relaxation sweeps. The dotted lines indicate the previous budget status. The shaded regions indicate the budget after relaxation. As described above, relaxation eliminates the high-frequency error components, but at the same time increases the component of the error in $\mathcal{R}(I_{2h}^h)$.

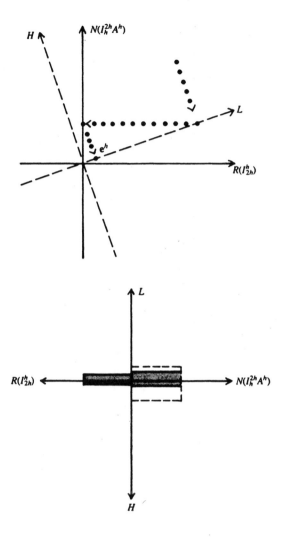

Figure 5.11: *The top diagram shows the effect of several more fine-grid relaxations on the error. Once again, all high-frequency components are removed as indicated by a projection onto the L-axis. The lower diagram shows the corresponding changes in the energy budget.*

The lower diagram of Fig. 5.10 shows the energy budget after a correction step. As we saw earlier, this step eliminates all of the error in $\mathcal{R}(I_{2h}^{h})$. Finally, further relaxation (Fig. 5.11) reduces the remaining oscillatory error modes, but reintroduces error with a small magnitude into $\mathcal{R}(I_{2h}^{h})$. A continuation of this cycle will further reduce the energy of the error; this reduction signifies convergence.

The subspace iteration diagrams, together with the energy budget diagrams, illustrate the way in which relaxation and correction work together. These two processes complement each other remarkably. When they are applied in tandem, the result is an extremely effective algorithm.

It should be remembered that this entire discussion has dealt only with a two-grid scheme. The V-cycle uses this scheme at all levels, ensuring that relaxation is always directed at the oscillatory modes of the current grid. As we have seen, two-grid correction without relaxation takes care of the smooth error components on the current fine grid. Therefore, by adding relaxation at all levels, all components of the error are eventually acted upon and quickly removed.

The efficiency of the two-grid scheme is further amplified by the FMG method. This scheme uses nested V-cycles to compute accurate initial guesses on coarser grids before relaxing on the finer grids. This ensures that a particular coarse-grid problem is solved to the level of discretization error before the more expensive fine-grid relaxations are begun. But even in these more elaborate procedures, it is the combination of relaxation and correction that provides the underlying power. Therefore, the arguments of this chapter, although somewhat qualitative, hold the key to the remarkable effectiveness of multigrid.

Exercises

1. **Energy norm.** Assume A is symmetric positive definite. As defined in this chapter, the A-energy inner product and the A-energy norm are given by

$$(\mathbf{u}, \mathbf{v})_A = (A\mathbf{u}, \mathbf{v}) \quad \text{and} \quad \|\mathbf{u}\|_A^2 = (\mathbf{u}, \mathbf{u})_A.$$

   (a) Show that these are acceptable definitions for an inner product and a norm.

   (b) Show that $\|\mathbf{r}\|_2 = \|\mathbf{e}\|_{A^2}$.

   (c) The error norm $\|\mathbf{e}\|_2$ is generally not computable. Is $\|\mathbf{e}\|_A$ computable? Is $\|\mathbf{e}\|_{A^2}$ computable?

2. **FMG error analysis.** A key step in the FMG error analysis is showing that

$$\|I_{2h}^h \mathbf{u}^{2h} - I_{2h}^h \mathbf{v}^{2h}\|_{A^h} = \|\mathbf{u}^{2h} - \mathbf{v}^{2h}\|_{A^{2h}}.$$

   Use the Galerkin property and the property of inner products that

$$(B\mathbf{u}, \mathbf{v}) = (\mathbf{u}, B^T \mathbf{v})$$

   to prove this equality for any two coarse-grid vectors.

3. **Discretization error.** The goal of this exercise is to consider the model problem in one dimension and show carefully that the discrete L^2 norm of the discretization error is bounded by a constant times h^2 (see equation (5.2)). Consider the boundary value problem

$$\begin{aligned} -u''(x) &= f(x) \quad \text{on} \quad (0,1), \\ u(0) = u(1) &= 0. \end{aligned}$$

   The discretized form of the problem is

$$\begin{aligned} (A^h \mathbf{u}^h)_i = -\frac{u_{i+1}^h - 2u_i^h + u_{i-1}^h}{h^2} &= f(x_i), \quad 1 \le i \le n, \\ u_0^h = u_{n+1}^h &= 0. \end{aligned}$$

(a) Let **u** be the vector consisting of values of the exact solution u sampled at the grid points; that is, $\mathbf{u}_i = u(x_i)$. Expanding $u(x_{i\pm1})$ in a Taylor series about x_i, show that

$$(A^h\mathbf{u})_i = -u''(x_i) - \frac{h^2}{24}(u^{(iv)}(\xi^+) + u^{(iv)}(\xi^-)),$$

where $x_{i-1} < \xi^- < x_i$ and $x_i < \xi^+ < x_{i+1}$.

(b) Assuming continuity of $u^{(iv)}$ and using the Intermediate Value Theorem, show that the *truncation error* is given by

$$\tau_i^h \equiv f(x_i) - (A^h\mathbf{u})_i = \frac{h^2}{24}(u^{(iv)}(\xi^+) + u^{(iv)}(\xi^-)) = \frac{h^2}{12}f''(\xi_i),$$

where $x_{i-1} < \xi_i < x_{i+1}$.

(c) Recall that if A is symmetric, then $\|A\|_2 = \rho(A)$, and that if $A\mathbf{w} = \lambda\mathbf{w}$, then $A^{-1}\mathbf{w} = \lambda^{-1}\mathbf{w}$. Use the eigenvalues of A^h and the definition of matrix norms to show that $\|(A^h)^{-1}\|_h \leq \gamma$, where $\gamma \approx \pi^{-2}$. Note that this bound is independent of h.

(d) Show that if f'' is bounded on $[0,1]$ and $v_i = f''(\xi_i)$, then $\|\mathbf{v}\|_h$ is bounded. Is this result true for the (unscaled) Euclidean norm?

(e) Combine the above facts and use properties of matrix norms to conclude that the discrete L^2 norm of the discretization error is bounded by a constant times h^2; begin by making the observation that

$$\|E^h\|_h = \|\mathbf{u}^h - \mathbf{u}\|_h = \|(A^h)^{-1}A^h(\mathbf{u}^h - \mathbf{u})\|_h.$$

4. **Effect of full weighting.** Verify that $I_h^{2h}\mathbf{w}_k^h = \cos^2\left(\frac{k\pi}{2n}\right)\mathbf{w}_k^{2h}$, where $w_{k,j}^h = \sin\left(\frac{jk\pi}{n}\right)$, $1 \leq k < \frac{n}{2}$, and I_h^{2h} is the full weighting operator.

5. **Effect of full weighting.** Verify that $I_h^{2h}\mathbf{w}_{k'}^h = -\sin^2\left(\frac{k\pi}{2n}\right)\mathbf{w}_k^{2h}$, where $k' = n - k$, $1 \leq k < \frac{n}{2}$, and I_h^{2h} is the full weighting operator. What happens to $\mathbf{w}_{n/2}^h$ under full weighting?

6. **Complementary modes.** Show that the complementary modes $\{\mathbf{w}_k^h, \mathbf{w}_{k'}^h\}$ on Ω^h are related by $w_{k',j}^h = (-1)^{j+1}w_{k,j}^h$, where $k' = n - k$.

7. **Null space of full weighting.** Verify that the vectors $\mathbf{n}_j = A^h\hat{\mathbf{e}}_j^h$, where j is odd and $\hat{\mathbf{e}}_j^h$ is the jth unit vector on Ω^h, form a basis for the null space of I_h^{2h}.

8. **Effect of linear interpolation.** Show that $I_{2h}^h\mathbf{w}_k^{2h} = c_k\mathbf{w}_k^h - s_k\mathbf{w}_{k'}^h$, where $1 \leq k < \frac{n}{2}$, $k' = n - k$, $c_k = \cos^2\left(\frac{k\pi}{2n}\right)$, $s_k = \sin^2\left(\frac{k\pi}{2n}\right)$, and I_{2h}^h is the linear interpolation operator.

9. **Two-grid correction scheme.** Using the previous results concerning the effects of A^h, I_h^{2h}, and I_{2h}^h on the fundamental modes, determine the effect of the coarse-grid correction operator without relaxation, $TG = I - I_{2h}^h(A^{2h})^{-1}I_h^{2h}A^h$, on the modes $\mathbf{w}_k$ and $\mathbf{w}_{k'}$, where $1 \leq k \leq \frac{n}{2}$, $k' = n - k$. (Recall that if $A\mathbf{w} = \lambda\mathbf{w}$, then $A^{-1}\mathbf{w} = \lambda^{-1}\mathbf{w}$.)

10. **Effect of two-grid correction.** Let R be an iteration matrix with the same eigenvectors, $\mathbf{w}_k$, as A^h and with eigenvalues λ_k. Consider the coarse-grid correction operator with ν sweeps of relaxation, $TG = [I - I_{2h}^h (A^{2h})^{-1} I_h^{2h} A^h] R^\nu$.

   (a) Determine how TG acts on the modes $\mathbf{w}_k$ and $\mathbf{w}_{k'}$, where $1 \le k \le \frac{n}{2}$, $k' = n - k$.

   (b) Find the specific convergence factors for the coarse-grid correction operator that uses one sweep of damped Jacobi with $\omega = \frac{2}{3}$.

11. **An important equivalence.** Prove that the null space of TG is the range of interpolation, where TG is the coarse grid correction operator without relaxation. Equation (5.12) of the text can be used to show that $N(TG) \supset \mathcal{R}(I_{2h}^h)$. Use equation (5.13) and the dimensions of the subspaces to show that $N(TG) = \mathcal{R}(I_{2h}^h)$.

12. **Two-dimensional five-point stencil.** When the two-dimensional model problem is discretized on a uniform grid with $h_x = h_y = h$, the coefficients at each grid point are given by the five-point stencil

$$A^h = \frac{1}{h^2} \begin{pmatrix} & -1 & \\ -1 & 4 & -1 \\ & -1 & \end{pmatrix}.$$

What does the stencil for $A^{2h} = I_h^{2h} A^h I_{2h}^h$ look like if I_{2h}^h is based on bilinear interpolation and I_h^{2h} is based on (a) full weighting? (b) injection?

13. **Two-dimensional nine-point stencil.** Repeat the previous problem with the nine-point stencil

$$A^h = \frac{1}{3h^2} \begin{pmatrix} -1 & -1 & -1 \\ -1 & 8 & -1 \\ -1 & -1 & -1 \end{pmatrix}.$$

Chapter 6

Nonlinear Problems

The preceding chapters presented multigrid methods applied to linear problems. Because of their iterative nature, multigrid ideas should be effective on nonlinear problems, where some form of iteration is usually imperative. Just as multigrid methods were developed to improve classical linear relaxation methods, we will see that they can also guide us in the use of nonlinear methods.

We begin by clarifying the most significant formal difference between linear and nonlinear systems of equations. Consider a system of nonlinear algebraic equations, $A(\mathbf{u}) = \mathbf{f}$, where $\mathbf{u}, \mathbf{f} \in \mathbf{R}^n$. (The notation $A(\mathbf{u})$, rather than $A\mathbf{u}$, signifies that the operator is nonlinear.) Suppose that $\mathbf{v}$ is an approximation to the exact solution $\mathbf{u}$. It is possible to define the error and the residual for this problem, much as we did earlier: the error is simply $\mathbf{e} = \mathbf{u} - \mathbf{v}$ and the residual is $\mathbf{r} = \mathbf{f} - A(\mathbf{v})$. If we now subtract the original equation, $A(\mathbf{u}) = \mathbf{f}$, from the definition of the residual, we find that

$$A(\mathbf{u}) - A(\mathbf{v}) = \mathbf{r}. \tag{6.1}$$

We are one step from the residual equation that we derived for linear systems in Chapter 2. However, because A is nonlinear, even though $\mathbf{u} - \mathbf{v} = \mathbf{e}$, we *cannot* conclude that $A(\mathbf{u}) - A(\mathbf{v}) = A(\mathbf{e})$. This means that we no longer have a simple linear residual equation, which necessitates changes in the methods that we devise for nonlinear problems. We must now use (6.1) as the residual equation.

It makes sense to review quickly a classical relaxation method for nonlinear systems; it will have a part to play in multigrid methods. Working with the system of nonlinear algebraic equations $A(\mathbf{u}) = \mathbf{f}$, one of the most frequently used methods is *nonlinear Gauss–Seidel* relaxation [16]. Suppose the jth equation of the system can be solved for the jth variable in terms of the other variables $(u_1, u_2, \ldots, u_{j-1}, u_{j+1}, \ldots, u_n)$. We write the result in the abstract form

$$u_j = M_j(u_1, u_2, \ldots, u_{j-1}, u_{j+1}, \ldots, u_n), \quad 1 \leq j \leq n.$$

Just as with linear systems, Gauss–Seidel relaxation takes the form

$$v_j \leftarrow M_j(v_1, v_2, \ldots, v_{j-1}, v_{j+1}, \ldots, v_n), \quad 1 \leq j \leq n,$$

where the current value of each component is used on the right side. For cases where this iteration cannot be formed explicitly, we use the same characterization that was used for linear Gauss–Seidel: for each $j = 1, 2, \ldots, n$, set the jth component of the residual to zero and solve for v_j. Because the residual is $\mathbf{r} = \mathbf{f} - A(\mathbf{v})$, this characterization amounts to solving $(A(\mathbf{v}))_j = \mathbf{f}_j$ for v_j. Equivalently, as in the linear case, we can let ϵ_j be the jth unit vector, so that the jth step of nonlinear Gauss–Seidel amounts to finding an $s \in \mathbf{R}$ such that

$$(A(\mathbf{v} + s\epsilon_j))_j = f_j, \quad 1 \le j \le n.$$

This is generally a nonlinear scalar equation in the scalar s, which can be solved efficiently with one or two steps of (scalar) Newton's method [16]. When an approximate solution s is found, the jth component is updated by $\mathbf{v} \leftarrow \mathbf{v} + s\epsilon_j$. Updating all n components sequentially in this way constitutes one iteration sweep of nonlinear Gauss–Seidel.

Now, to see how residual equation (6.1) can be used as a basis for multigrid methods, let $\mathbf{v}$ be the current approximation and replace the exact solution $\mathbf{u}$ by $\mathbf{v} + \mathbf{e}$. Then residual equation (6.1) becomes

$$A(\mathbf{v} + \mathbf{e}) - A(\mathbf{v}) = \mathbf{r}. \tag{6.2}$$

Expanding $A(\mathbf{v} + \mathbf{e})$ in a Taylor series (in n variables) about $\mathbf{v}$ and truncating the series after two terms, we have the linear system of equations

$$J(\mathbf{v})\mathbf{e} = \mathbf{r}, \tag{6.3}$$

where $J = (\partial A_i / \partial u_j)$ is the $n \times n$ Jacobian matrix. Linear system (6.3) represents an approximation to nonlinear system (6.1). It can be solved for $\mathbf{e}$ and the current approximation $\mathbf{v}$ can be updated by $\mathbf{v} \leftarrow \mathbf{v} + \mathbf{e}$. Iteration of this step is *Newton's method*. But how should linear system (6.3) be solved at each step? One highly recommended option is to use multigrid. Such a combination of Newton's method for the outer iteration and multigrid for the (linear) inner iteration is called *Newton-multigrid*.

Newton-multigrid can be effective, but it does not use multigrid ideas to treat the nonlinearity directly. To do this, we return to the residual equation,

$$A(\mathbf{v} + \mathbf{e}) - A(\mathbf{v}) = \mathbf{r},$$

and use it in the familiar two-grid setting. Suppose we have found an approximation, $\mathbf{v}^h$, to the original fine-grid problem

$$A^h(\mathbf{u}^h) = \mathbf{f}^h. \tag{6.4}$$

Proceeding as we did for the linear problem, we now want to use the residual equation on the coarse grid Ω^{2h} to approximate $\mathbf{e}^h$, the error in $\mathbf{v}^h$. Using the above argument, the residual equation on the coarse grid appears as

$$A^{2h}(\mathbf{v}^{2h} + \mathbf{e}^{2h}) - A^{2h}(\mathbf{v}^{2h}) = \mathbf{r}^{2h}, \tag{6.5}$$

where A^{2h} denotes the coarse-grid operator, $\mathbf{r}^{2h}$ is the coarse-grid residual, $\mathbf{v}^{2h}$ is a coarse-grid approximation to $\mathbf{v}^h$, and $\mathbf{e}^{2h}$ is a coarse-grid approximation to $\mathbf{e}^h$. Once $\mathbf{e}^{2h}$ is computed, the fine-grid approximation can be updated by $v^h \leftarrow v^h + I_{2h}^h e^{2h}$.

We have already encountered the coarse-grid residual $\mathbf{r}^{2h}$. Nothing changes here; we simply choose it to be the restriction of the fine-grid residual to the coarse grid:

$$\mathbf{r}^{2h} = I_h^{2h} \mathbf{r}^h = I_h^{2h}(\mathbf{f}^h - A^h(\mathbf{v}^h)).$$

Newton's Method. Perhaps the best known and most important method for solving nonlinear equations is *Newton's method*, which can be derived as follows. Suppose we wish to solve the scalar equation $F(x) = 0$. We expand F in a Taylor series about an initial guess x:

$$F(x + s) = F(x) + sF'(x) + \frac{s^2}{2}F''(\xi),$$

where ξ is between x and $x + s$. If $x + s$ is the solution, then (neglecting the higher-order terms) the series becomes $0 = F(x) + sF'(x)$, from which $s = -F(x)/F'(x)$. Thus, we can update the initial guess using $x \leftarrow x - F(x)/F'(x)$. Newton's method results by iterating this process:

$$x \leftarrow x - \frac{F(x)}{F'(x)}.$$

Newton's method for a system of n nonlinear equations is a straightforward extension of the scalar Newton's method. We write the system in vector form as

$$\mathbf{F}(\mathbf{x}) \equiv \begin{bmatrix} f_1(x_1, x_2, \ldots, x_n) \\ f_2(x_1, x_2, \ldots, x_n) \\ \vdots \\ f_n(x_1, x_2, \ldots, x_n) \end{bmatrix} = \begin{bmatrix} 0 \\ 0 \\ \vdots \\ 0 \end{bmatrix}.$$

Expanding in a Taylor series yields

$$\mathbf{F}(\mathbf{x} + \mathbf{s}) = \mathbf{F}(\mathbf{x}) + J(\mathbf{x})\mathbf{s} + \text{higher-order terms},$$

where $J(\mathbf{x})$ is the Jacobian matrix

$$J(\mathbf{x}) = \begin{bmatrix} \frac{\partial f_1}{\partial x_1} & \frac{\partial f_1}{\partial x_2} & \cdots & \frac{\partial f_1}{\partial x_n} \\ \frac{\partial f_2}{\partial x_1} & \frac{\partial f_2}{\partial x_2} & \cdots & \frac{\partial f_2}{\partial x_n} \\ \vdots & \vdots & \ddots & \vdots \\ \frac{\partial f_n}{\partial x_1} & \frac{\partial f_n}{\partial x_2} & \cdots & \frac{\partial f_n}{\partial x_n} \end{bmatrix}.$$

Newton's method results when we replace $\mathbf{F}(\mathbf{x} + \mathbf{s}) = \mathbf{0}$ by $\mathbf{F}(\mathbf{x}) + J(\mathbf{x})\mathbf{s} = \mathbf{0}$. Solving for the vector $\mathbf{s}$ yields the iteration

$$\mathbf{x} \leftarrow \mathbf{x} - [J(\mathbf{x})]^{-1}\mathbf{F}(\mathbf{x}).$$

But what about the current approximation $\mathbf{v}^{2h}$? When we move to the coarse grid, it makes sense to restrict the current fine-grid approximation to the coarse grid, often using the same transfer operator used for the residual: $\mathbf{v}^{2h} = I_h^{2h}\mathbf{v}^h$.

Making these substitutions in the residual equation yields

$$A^{2h}(\underbrace{I_h^{2h}\mathbf{v}^h + \mathbf{e}^{2h}}_{\mathbf{u}^{2h}}) = A^{2h}(I_h^{2h}\mathbf{v}^h) + I_h^{2h}(\mathbf{f}^h - A^h(\mathbf{v}^h)). \tag{6.6}$$

The right side of this nonlinear system is known. The goal is to find or approximate the solution to this system, which we have denoted $\mathbf{u}^{2h}$. The coarse-grid error approximation, $\mathbf{e}^{2h} = \mathbf{u}^{2h} - I_h^{2h}\mathbf{v}^h$, can then be interpolated up to the fine grid and used to correct the fine-grid approximation $\mathbf{v}^h$. This correction step takes the form

$$\mathbf{v}^h \leftarrow \mathbf{v}^h + I_{2h}^h\mathbf{e}^{2h} \quad \text{or} \quad \mathbf{v}^h \leftarrow \mathbf{v}^h + I_{2h}^h(\mathbf{u}^{2h} - I_h^{2h}\mathbf{v}^h).$$

The scheme we have just outlined is the most commonly used nonlinear version of multigrid. It is called the *full approximation scheme* (FAS) because the coarse-grid problem is solved for the full approximation $\mathbf{u}^{2h} = I_h^{2h}\mathbf{v}^h + \mathbf{e}^{2h}$ rather than the error $\mathbf{e}^{2h}$. A two-grid version of this scheme is described as follows:

Full Approximation Scheme (FAS)

- Restrict the current approximation and its fine-grid residual to the coarse grid: $\mathbf{r}^{2h} = I_h^{2h}(\mathbf{f}^h - A^h(\mathbf{v}^h))$ and $\mathbf{v}^{2h} = I_h^{2h}\mathbf{v}^h$.

- Solve the coarse-grid problem $A^{2h}(\mathbf{u}^{2h}) = A^{2h}(\mathbf{v}^{2h}) + \mathbf{r}^{2h}$.

- Compute the coarse-grid approximation to the error: $\mathbf{e}^{2h} = \mathbf{u}^{2h} - \mathbf{v}^{2h}$.

- Interpolate the error approximation up to the fine grid and correct the current fine-grid approximation: $\mathbf{v}^h \leftarrow \mathbf{v}^h + I_{2h}^h\mathbf{e}^{2h}$.

There are several observations to be made about this method. It is worth noting that if A is a linear operator, then the FAS scheme reduces directly to the (linear) two-grid correction scheme (Exercise 1). Thus, FAS can be viewed as a generalization of the two-grid correction scheme to nonlinear problems.

Less obvious is the fact that an exact solution of the fine-grid problem is a fixed point of the FAS iteration (Exercise 2). This fixed point property, which is a desirable attribute of most iterative methods, means that the process stalls at the exact solution.

A third observation is that the FAS coarse-grid equation can be written as

$$A^{2h}(\mathbf{u}^{2h}) = \mathbf{f}^{2h} + \boldsymbol{\tau}_h^{2h},$$

where the *tau correction* $\boldsymbol{\tau}_h^{2h}$ is defined by

$$\boldsymbol{\tau}_h^{2h} = A^{2h}(I_h^{2h}\mathbf{v}^h) - I_h^{2h}A^h(\mathbf{v}^h).$$

One of the many consequences of this relationship is that the solution of the coarse-grid FAS equation, $\mathbf{u}^{2h}$, is not the same as the solution of the original coarse-grid equation $A^{2h}(\mathbf{u}^{2h}) = \mathbf{f}^{2h}$ because $\boldsymbol{\tau}_h^{2h} \neq \mathbf{0}$ generally. In fact, as FAS processing advances, $\mathbf{u}^{2h}$ begins to achieve *accuracy* that compares to that of the solution on the finest grid, albeit at the *resolution* of grid $2h$. This tau correction relationship allows us to view FAS as a way to alter the coarse-grid equations so that their approximation properties are substantially enhanced.

Here is another important observation. Because the second step in the above procedure involves a nonlinear problem itself, FAS involves an inner and an outer iteration; the outer iteration is the FAS correction scheme, while the inner iteration is usually a standard relaxation method such as nonlinear Gauss–Seidel. A true multilevel FAS process would be done recursively by approximating solutions to the Ω^{2h} problem using the next coarsest grid, Ω^{4h}. Thus, FAS, like its linear counterparts, is usually implemented as a V-cycle or W-cycle scheme.

In earlier chapters, we saw the importance of full multigrid (FMG) for obtaining a good initial guess for the fine-grid problem. The convergence of nonlinear iterations depends even more critically on a good initial guess. Typically, the better the initial guess used on the fine grid, the more linear the fine-grid problem appears, and the more effective the fine-grid solver will be. When FMG is used for nonlinear problems, the interpolant $I_{2h}^h \mathbf{u}^{2h}$ is generally accurate enough to be in the basin of attraction of the fine-grid solver. Thus, whether we use Newton-multigrid or FAS V-cycles on each new level, we can expect one FMG cycle to provide accuracy to the level of discretization, unless perhaps the nonlinearity is exceptionally strong. These options are investigated in a numerical example later in the chapter.

Example: An algebraic problem. We need to warn the reader that this example should not be taken too seriously: it was devised only as a way to illustrate the *mechanics* of the FAS scheme. The problem does not really need the power of multigrid because it can be effectively treated by a good classical relaxation scheme. More importantly, as with most nonlinear problems, the example has subtleties that must be addressed before solution techniques should even be considered (Exercise 4).

Consider the following nonlinear algebraic system of n equations and n unknowns $\mathbf{u}^h \in \mathbf{R}^n$:

$$A_j^h(\mathbf{u}^h) \equiv u_j^h \, u_{j+1}^h = f_j^h, \qquad 1 \le j \le n, \tag{6.7}$$

where the vector $\mathbf{f}^h \in \mathbf{R}^n$ is given. (The notation $A_j^h(\mathbf{u}^h)$ means the jth component of $A^h(\mathbf{u}^h)$.) We impose the periodic boundary condition $u_{n+1}^h = u_1^h$ to close the problem. Suppose we have a fine-grid approximation, $\mathbf{v}^h$, obtained by relaxation. In passing, note that Gauss–Seidel relaxation is easy to formulate for this problem: it consists of the replacement steps

$$v_j^h \leftarrow \frac{f_j^h}{v_{j+1}^h}, \quad 1 \le j \le n. \tag{6.8}$$

To see what the FAS correction looks like, assume that n is a positive even integer and write coarse-grid residual equation (6.5) as

$$A_j^{2h}(\mathbf{v}^{2h} + \mathbf{e}^{2h}) - A_j^{2h}(\mathbf{v}^{2h}) = r_j^{2h}, \quad 1 \le j \le \frac{n}{2}.$$

In component form, these equations are

$$
\begin{aligned}
(v_j^{2h} + e_j^{2h})(v_{j+1}^{2h} + e_{j+1}^{2h}) - v_j^{2h} v_{j+1}^{2h} &= r_j^{2h} \quad \text{or} \\
v_j^{2h} e_{j+1}^{2h} + v_{j+1}^{2h} e_j^{2h} + e_j^{2h} e_{j+1}^{2h} &= r_j^{2h}, \quad 1 \le j \le \frac{n}{2}.
\end{aligned}
$$

Notice how the term $v_j^{2h} v_{j+1}^{2h}$ cancels. Now, e_j^{2h} and e_{j+1}^{2h} are the unknowns in the jth equation; the coefficients v_j^{2h}, v_{j+1}^{2h}, and r_j^{2h} must come from the fine grid. We can obtain the terms v_j^{2h} and v_{j+1}^{2h} from the fine grid by injection:

$$v_j^{2h} = I_h^{2h} v_{2j}^h = v_{2j}^h, \quad v_{j+1}^{2h} = I_h^{2h} v_{2j+2}^h = v_{2j+2}^h.$$

Similarly, the coarse-grid residual (using injection) is given by

$$r_j^{2h} = I_h^{2h} r_{2j}^h = f_{2j}^h - v_{2j}^h v_{2j+1}^h.$$

The equations that must be solved on the coarse grid for e_j^{2h} are therefore given by

$$v_{2j}^h e_{j+1}^{2h} + v_{2j+2}^h e_j^{2h} + e_j^{2h} e_{j+1}^{2h} = f_{2j}^h - v_{2j}^h v_{2j+1}^h, \quad 1 \le j \le \frac{n}{2}.$$

This system corresponds to the general FAS equation (6.6). In this example, with a specific form for the operator A^h, we have simplified the right side of (6.6), resulting in an explicit system for e_j^{2h}. This system would typically be handled by a standard relaxation method. Note that Gauss–Seidel is again fairly simple because the jth equation is linear in e_j^{2h}, although the *system* of equations is nonlinear in the vector $\mathbf{e}^h$. In any case, after approximations to e_j^{2h} have been computed, they can then be interpolated up to the fine grid to correct $\mathbf{v}^h$.   ∞

Example: Formulating FAS for a boundary value problem. Consider the two-point boundary value problem

$$\begin{aligned}
-u''(x) + \gamma u(x)u'(x) &= f(x), \quad 0 < x < 1, \quad (6.9) \\
u(0) = u(1) &= 0.
\end{aligned}$$

The source term f and a constant $\gamma > 0$ are given. To set up the FAS solution to this problem, let Ω^h consist of the grid points $x_j = \frac{j}{n}$, for some positive even integer n, and let $u_j = u(x_j)$ and $f_j = f(x_j)$, for $j = 0, 1, \ldots, n$. Of the many possible ways to discretize this differential equation, consider the following finite difference scheme:

$$A_j^h(\mathbf{u}) = \frac{-u_{j-1}^h + 2u_j^h - u_{j+1}^h}{h^2} + \gamma u_j^h \left(\frac{u_{j+1}^h - u_{j-1}^h}{2h} \right) = f_j, \quad 1 \le j \le n-1.$$

Note the use of a centered difference approximation to $u'(x_j)$. One consequence of this choice is that this *scalar* equation is linear in u_j^h even though the *vector* equation $A^h(\mathbf{u}^h) = \mathbf{f}^h$ is nonlinear in $\mathbf{u}^h$. Thus, nonlinear Gauss–Seidel requires no Newton step and can be stated explicitly as follows:

$$u_j^h \leftarrow 2 \frac{h^2 f_j + \left(u_{j-1}^h + u_{j+1}^h \right)}{4 + h\gamma(u_{j+1}^h - u_{j-1}^h)}, \quad 1 \le j \le n-1. \quad (6.10)$$

To examine the FAS correction step, assume that an approximation $\mathbf{v}^h$ has been obtained on the fine grid. Residual equation (6.5), given by

$$A_j^{2h}(\mathbf{v}^{2h} + \mathbf{e}^{2h}) - A_j^{2h}(\mathbf{v}^{2h}) = r_j^{2h},$$

appears in component form as

$$\frac{-(v_{j-1}^{2h} + e_{j-1}^{2h}) + 2(v_j^{2h} + e_j^{2h}) - (v_{j+1}^{2h} + e_{j+1}^{2h})}{4h^2}$$

$$+ \gamma(v_j^{2h} + e_j^{2h}) \left(\frac{v_{j+1}^{2h} - v_{j-1}^{2h}}{4h} + \frac{e_{j+1}^{2h} - e_{j-1}^{2h}}{4h} \right) - \frac{-v_{j-1}^{2h} + 2v_j^{2h} - v_{j+1}^{2h}}{4h^2}$$

$$- \gamma v_j^{2h} \left(\frac{v_{j+1}^{2h} - v_{j-1}^{2h}}{4h} \right) = r_j^{2h}, \quad 1 \le j \le \frac{n}{2} - 1.$$

Canceling terms leaves the coarse-grid equation for the unknowns e_j^{2h}:

$$\frac{-e_{j-1}^{2h} + 2e_j^{2h} - e_{j+1}^{2h}}{4h^2} + \gamma v_j^{2h}\left(\frac{e_{j+1}^{2h} - e_{j-1}^{2h}}{4h}\right) + \gamma e_j^{2h}\left(\frac{v_{j+1}^{2h} - v_{j-1}^{2h}}{4h}\right)$$

$$+ \gamma e_j^{2h}\left(\frac{e_{j+1}^{2h} - e_{j-1}^{2h}}{4h}\right) = \underbrace{I_h^{2h}(f_j^h - A_j^h(v^h))}_{r_j^{2h}}.$$

As before, the terms $v_j^{2h}, v_{j+1}^{2h}, v_{j-1}^{2h}$, and r_j^{2h} are obtained by restriction from the fine grid.

This equation is the analogue of FAS equation (6.6), although we have written out the terms explicitly. Approximations to the solution, e_j^{2h}, of this equation must be computed using a relaxation method such as nonlinear Gauss–Seidel (which can again be carried out explicitly). These corrections are interpolated up to the fine grid and used to update the fine-grid approximation $\mathbf{v}^h$. ∞

Numerical example: FAS for a boundary value problem. To study the performance of FAS on boundary value problem (6.9), we choose the exact solution

$$u(x) = e^x(x - x^2),$$

which results in the source term

$$f(x) = (x^2 + 3x)e^x + \gamma(x^4 - 2x^2 + x)e^{2x}.$$

The problem is discretized on a grid with $n = 512$ (511 interior points). On the coarsest grid, consisting of one interior point and two boundary points, the problem is solved exactly by

$$u_1 = \frac{f_1}{8}.$$

In this experiment, we treat the problem using full weighting, linear interpolation, and a (2,1) FAS V-cycle based on nonlinear Gauss–Seidel (6.10). For $\gamma = 0$, the problem reduces to the one-dimensional Poisson equation, which is the standard model problem for multigrid. In this instance, the FAS V-cycle reduces to the standard V-cycle. As γ increases, the nonlinear term $\gamma u(x)u'(x)$ begins to dominate the problem.

Table 6.1 displays the performance of FAS for various values of γ. Starting with an initial guess of $\mathbf{v}^h = \mathbf{0}$, the FAS V-cycles are carried out until the norm of the residual vector is less than 10^{-10}. Displayed for each choice of γ is the number of FAS V-cycles required to achieve this tolerance, as well as the average convergence factor per cycle. For $\gamma \leq 10$, FAS performance is essentially the same as for the linear case, $\gamma = 0$. As γ increases, the performance degrades slowly until, for $\gamma \geq 40$, FAS no longer converges.

We see similar qualitative behavior if we change the exact solution to $u(x) = x - x^2$ so that $f(x) = 2 + \gamma(x - x^2)(1 - 2x)$. FAS performance for this problem is summarized in Table 6.2. Now the method converges for $\gamma < 100$, although performance degrades more noticeably as γ approaches $\gamma = 100$. For $\gamma \geq 100$, the method no longer converges. FAS is an effective solver for this problem, even for cases where the nonlinear term is relatively large.

	γ						
	0	1	10	25	35	39	≥ 40
Convergence factor	.096	.096	.096	.163	.200	.145	NC
Number of FAS cycles	11	11	11	14	16	13	NC

Table 6.1: *FAS performance for the problem* $-u'' + \gamma\,u\,u' = f$, *with the exact solution* $u(x) = e^x(x - x^2)$, *discretized on a grid with* 511 *interior points. Shown are average convergence factors and the number of FAS cycles needed to converge from a zero initial guess to a residual norm of* 10^{-10}. *NC indicates that the method did not converge.*

	γ						
	0	1	10	50	70	90	≥ 100
Convergence factor	.096	.096	.094	.148	.432	.728	NC
Number of FAS V-cycles	11	11	11	13	30	79	NC

Table 6.2: *FAS performance for the problem* $-u'' + \gamma\,u\,u' = f$, *with the exact solution* $u(x) = x - x^2$, *discretized on a grid with* 511 *interior points. Shown are average convergence factors and the number of FAC cycles needed to converge from a zero initial guess to a residual norm of* 10^{-10}. *NC indicates that the method did not converge.* ◇◇

Numerical example: Two-dimensional boundary value problem. We finish the numerical examples with a study of the performance of FAS and Newton solvers applied to the two-dimensional nonlinear problem

$$-\Delta u(x,y) + \gamma\,u(x,y)\,e^{u(x,y)} = f(x,y) \quad \text{in } \Omega, \quad (6.11)$$
$$u(x,y) = 0 \quad \text{on } \partial\Omega, \quad (6.12)$$

where Ω is the unit square $[0,1] \times [0,1]$. For $\gamma = 0$, this problem reduces to the model problem. We discretize this equation on uniform grids in both directions, with grid spacings of $h = \frac{1}{64}$ and $h = \frac{1}{128}$. Using the usual finite difference operator, the equation for the unknown $u_{i,j}$ at $(x_i, y_j) = (ih, jh)$ becomes

$$\frac{4u_{i,j} - u_{i-1,j} - u_{i+1,j} - u_{i,j-1} - u_{i,j+1}}{h^2} + \gamma\,u_{i,j}\,e^{u_{i,j}} = f_{i,j}, \quad 1 < i, j < n. \quad (6.13)$$

To enforce the boundary condition, we set

$$u_{0,j} = u_{N,j} = u_{i,0} = u_{i,N} = 0$$

wherever these terms appear in the equations.

We consider several approaches to solving this problem. An FAS solver can be implemented in a straightforward way. We use full weighting and linear interpolation for the intergrid transfer operators. For the coarse-grid versions of the nonlinear operator, we use discretization (6.13) with the appropriate grid spacing $(2h, 4h, \ldots)$. Because the individual component equations of the system are nonlinear, the nonlinear Gauss–Seidel iteration uses scalar Newton's method to solve the (i, j)th equation for $u_{i,j}$ (Exercise 5):

$$u_{i,j} \leftarrow u_{i,j} - \frac{h^{-2}(4u_{i,j} - u_{i-1,j} - u_{i+1,j} - u_{i,j-1} - u_{i,j+1}) + \gamma u_{i,j} e^{u_{i,j}} - f_{i,j}}{4h^{-2} + \gamma (1 + u_{i,j}) e^{u_{i,j}}}.$$

$$(6.14)$$

The nonlinear Gauss–Seidel smoother is not intended to solve the system exactly, so we need not solve for $u_{i,j}$ exactly. Although no fixed rule exists to determine how many Newton steps are required in each nonlinear Gauss–Seidel sweep, a small number usually suffices (we used one Newton step in this example). It should be noted that the number of Newton steps used on the scalar problem can have a significant impact on both the effectiveness and the cost of the FAS algorithm.

In the previous example, it was simple to solve the equation on the coarsest grid exactly. This is not the case here, where, on a 3×3 grid with a single interior point, the equation to be solved for $u_{1,1}$ is

$$16u_{1,1} + \gamma u_{1,1} e^{u_{1,1}} = f_{1,1}.$$

Because this equation is nonlinear, the "solve" on the coarse grid requires the use of Newton's Method. All experiments presented here use a 3×3 grid as the coarsest grid, and in all cases it was determined experimentally that a single Newton step was sufficient there.

In this example, we use the source term

$$f(x, y) = 2((x - x^2) + (y - y^2)) + \gamma (x - x^2)(y - y^2) e^{(x-x^2)(y-y^2)},$$

which corresponds to the exact solution

$$u(x, y) = (x - x^2)(y - y^2).$$

The first test examines the effect of FAS for various choices of γ, from the linear case ($\gamma = 0$) through cases in which the nonlinear part of the operator dominates ($\gamma = 10,000$). For each test, the problem is solved on a 127×127 interior grid. The problem is deemed solved when the norm of the residual is less than 10^{-10}, at which point the average convergence rate is reported. Also reported is the number of V-cycles required to achieve the desired tolerance. The results appear in Table 6.3.

	\multicolumn{6}{c}{γ}
	0	1	10	100	1000	10000
Convergence factor	.136	.135	.124	.098	.072	.039
Number of FAS cycles	12	12	11	11	10	8

Table 6.3: *FAS performance for the problem $-\Delta u + \gamma u e^u = f$, discretized on a grid with 127×127 interior points.*

One feature of interest is that this problem becomes *easier* to solve as the nonlinear term becomes more dominant. Note that this particular nonlinear term, ue^u, involves only the unknown $u_{i,j}$, and none of its neighboring values. Therefore, as the nonlinear term becomes more dominant, the problem becomes more local in nature. As this occurs, the smoothing steps, being local solvers, become more effective. Indeed, with enough dominance of the nonlinear term, the problem could be solved entirely with the nonlinear Gauss–Seidel smoother. This phenomenon is analogous to increasing diagonal dominance in the linear case. Of course, this type of behavior would not be expected for other types of nonlinearity.

It is useful to examine how the FAS solver performs compared to Newton's method applied to the nonlinear system. The Jacobian matrix for this problem is a block tridiagonal system

$$
J(\mathbf{u}) = \begin{bmatrix}
J_1 & B & & & & \\
B & J_2 & B & & & \\
 & B & J_3 & B & & \\
 & & \ddots & \ddots & \ddots & \\
 & & & B & J_{N-2} & B \\
 & & & & B & J_{N-1}
\end{bmatrix},
$$

where each of the block matrices is $(n-1) \times (n-1)$. The off-diagonal blocks B are all $-\frac{1}{h^2}$ times the identity matrix. The diagonal blocks are tridiagonal, with the constant value $-\frac{1}{h^2}$ on the super- and sub-diagonals. The diagonal entries of J_j, corresponding to the grid locations $(x_i, y_j) = (ih, jh)$ for fixed j and $1 \le i \le n-1$, are given by

$$
(J_j)_{i,i} = \frac{4}{h^2} + \gamma\, u_{i,j}\, e^{u_{i,j}}.
$$

To perform the Newton iteration, we choose two of the many ways to invert the Jacobian matrix. The first is to compute the LU decomposition of $J(\mathbf{u})$ and to use it to solve the system. With a lower triangular matrix L and an upper triangular matrix U such that $LU = J(\mathbf{u})$, we solve the system $J(\mathbf{u})\mathbf{s} = F(\mathbf{u})$, first solving $L\mathbf{y} = F(\mathbf{u})$ and then solving $U\mathbf{s} = \mathbf{y}$. Because $J(\mathbf{u})$ is sparse and has a narrow band of nonzero coefficients, the factors L and U are also sparse, narrow-banded matrices. Indeed, the banded LU decomposition can be computed quite efficiently; and because the factors L and U are triangular, solving for $\mathbf{y}$ and $\mathbf{s}$ is also fast. Table 6.4 gives the results of applying Newton's method, with the band LU solver, to the problem $-\Delta u + \gamma\, u\, e^u = f$, using a 63×63 interior grid.

	\multicolumn{6}{c}{γ}					
	0	1	10	100	1000	10000
Convergence factor	2.6e-13	3.9e-5	7.4e-5	3.2e-4	1.9e-4	1.2e-4
Number of Newton iterations	1	3	3	4	4	4

Table 6.4: *Performance of Newton's Method for the problem* $-\Delta u + \gamma\, u\, e^u = f$, *discretized on a 63×63 interior grid.*

A comparison of Tables 6.3 and 6.4 indicates that convergence of Newton's method is much faster than that of FAS. It is well known that Newton's method converges quadratically, so this is not surprising. To make this comparison useful, however, it is important to compare the computational costs of the methods. We present some empirical evidence on this question shortly.

First, however, we consider an alternative method for inverting the Jacobian system in the Newton iteration, namely, V-cycles. Certainly in the linear case

($\gamma = 0$), we know this is an effective method: both the original and the Jacobian problems are discrete Poisson equations. As γ increases, the diagonal dominance of the Jacobian improves, so we should expect even better performance.

If we solve the Jacobian system to the same level of accuracy by V-cycles as was done with the *LU* decomposition, the results should be essentially the same as in Table 6.4. However, it is probably much more efficient to use only a few V-cycles to solve the Jacobian system approximately. This technique is known as the *inexact Newton's method* [8], which we denote Newton-MG.

We now compare the performance of FAS, Newton's method, and Newton-MG in the following numerical experiment. Consider the same operator as before with a different source term:

$$-\Delta u + \gamma u e^u = \left(\left(9\pi^2 + \gamma e^{(x^2-x^3)\sin(3\pi y)}\right)(x^2 - x^3) + 6x - 2\right)\sin(3\pi y). \quad (6.15)$$

With $\gamma = 10$, the exact solution is $u(x,y) = (x^2 - x^3)\sin(3\pi y)$. This solution (as opposed to the previous polynomial solution) results in a nontrivial discretization error. The problem is discretized with $n = 128$ so that the interior grid is 127×127.

Method	No. outer iterations	No. inner iterations	Megaflops
Newton	3	–	1660.6
Newton-MG	3	20	56.4
Newton-MG	4	10	38.5
Newton-MG	5	5	25.1
Newton-MG	10	2	22.3
Newton-MG	19	1	24.6
FAS	11	–	27.1

Table 6.5: *Comparison of FAS, Newton, and Newton-multigrid methods for the problem $-\Delta u + \gamma\, u\, e^u = f$ on a 127×127 grid. In all cases, a zero initial guess is used.*

Table 6.5 shows the costs of FAS, Newton (with a direct solve), and Newton-MG applied to (6.15), resulting in the nonlinear system (6.13). The iteration is stopped when the residual norm is less than 10^{-10}. The column labeled *outer iterations* lists how many V-cycles (for FAS) or Newton steps (for Newton and Newton-MG) are required to achieve the desired tolerance. For Newton-MG, we varied the number of V-cycles used to solve the Jacobian system. The column labeled *inner iterations* gives the number of V-cycles used in the approximate inversion of the Jacobian system. The last column, labeled *Megaflops*, is the number of millions of floating-point operations required to achieve the desired tolerance (as determined by the MAT-LAB *flops* function). For this example, using these performance measurements, it appears that both the multigrid-based methods are much more efficient than Newton's method using a direct solver. Furthermore, Newton-MG compares well to FAS when the number of inner MG iterations is properly tuned.

These results should not be taken too seriously. While the convergence properties may be representative for a fairly broad class of nonlinear problems, the operation counts are likely to vary dramatically with the character of the nonlinearity, details of implementation, computer architecture, and programming language. It should also be remembered that we have not accounted for the cost of evaluating the nonlinear function, which is highly problem-dependent. Therefore, it may be risky to draw any general conclusions from this single experiment. On the other hand, it does seem fairly clear that the multigrid-based methods will outperform Newton's method using direct solvers when the problems are large. It is also clear that only a few V-cycles should be used in the Newton scheme if it is to be competitive with FAS. Finally, in the end, there may be very little difference in the performance between the two carefully designed multigrid-based schemes. The choice may depend largely on convenience and other available algorithm features (for example, τ-extrapolation in FAS [4]).

There is, of course, one further option we should consider: combining the Newton and FAS methods with an FMG scheme. The idea is to apply the basic solver (either Newton or FAS) on successively finer grids of the FMG cycle: for each new fine grid, the initial guess is obtained by first solving the nonlinear problem on the next coarser grid. In the linear case, we showed that convergence to the level of discretization was achieved in one FMG cycle. Table 6.6 shows the results of applying the FMG–FAS combination to (6.15); it suggests that we can hope for the same kind of performance in the nonlinear case.

The discrete L^2 norms are shown for the residual and error (difference between computed and sampled continuous solutions), after one FMG–FAS cycle and eight subsequent FAS V-cycles on the fine grid. Both the FMG cycle and the FAS V-cycles were performed using nonlinear Gauss–Seidel (2,1) sweeps. Observe that the norm of the error is reduced to 2.0×10^{-5} by the FMG–FAS cycle alone. Further FAS V-cycling does not reduce the error, indicating that it has reached the level of discretization error. However, subsequent FAS V-cycles do reduce the residual norm further until it reaches the prescribed tolerance of 10^{-10}. The column labeled *Mflops* in the table gives the cumulative number of floating-point operations after each stage of the computation.

Cycle	$\|\mathbf{r}^h\|_h$	Ratio	$\|\mathbf{e}\|_h$	Mflops
FMG–FAS	1.07e−2		2.00e−5	3.1
FAS V 1	6.81e−4	0.064	2.44e−5	5.4
FAS V 2	5.03e−5	0.074	2.49e−5	7.6
FAS V 3	3.89e−6	0.077	2.49e−5	9.9
FAS V 4	3.25e−7	0.083	2.49e−5	12.2
FAS V 5	2.98e−8	0.092	2.49e−5	14.4
FAS V 6	2.94e−9	0.099	2.49e−5	16.7
FAS V 7	3.01e−10	0.102	2.49e−5	18.9
FAS V 8	3.16e−11	0.105	2.49e−5	21.2

Table 6.6: *Performance of the FMG–FAS cycle, followed by eight FAS V-cycles, on* $-\Delta u + \gamma u e^u = f$, *with* $\gamma = 10$. *The grid size is* 127×127. *Note that one FMG–FAS cycle reduces the error to the level of discretization error, and that subsequent FAS V-cycles further reduce the residual norm quickly to the prescribed tolerance of* 10^{-10}.

We show analogous results in Table 6.7 for FMG with a Newton solver applied to (6.15). Here we again use FMG, applying one step of Newton-MG on each level in the FMG process. Each Newton step starts with an initial guess from the next coarser grid and uses one (2,1) V-cycle. The discrete L^2 norms are shown for the residual and error after one FMG–Newton-MG cycle followed by subsequent

Cycle	$\|\mathbf{r}^h\|_h$	Ratio	$\|\mathbf{e}\|_h$	Mflops
FMG–Newton-MG	1.06e−002		2.50e−005	2.4
Newton-MG 1	6.72e−004	0.063	2.49e−005	4.1
Newton-MG 2	5.12e−005	0.076	2.49e−005	5.8
Newton-MG 3	6.30e−006	0.123	2.49e−005	7.5
Newton-MG 4	1.68e−006	0.267	2.49e−005	9.2
Newton-MG 5	5.30e−007	0.315	2.49e−005	10.9
Newton-MG 6	1.69e−007	0.319	2.49e−005	12.6
Newton-MG 7	5.39e−008	0.319	2.49e−005	14.3
Newton-MG 8	1.72e−008	0.319	2.49e−005	16.0
Newton-MG 9	5.50e−009	0.319	2.49e−005	17.7
Newton-MG 10	1.76e−009	0.319	2.49e−005	19.4
Newton-MG 11	5.61e−010	0.319	2.49e−005	21.1
Newton-MG 12	1.79e−010	0.319	2.49e−005	22.8
Newton-MG 13	5.71e−011	0.319	2.49e−005	24.5

Table 6.7: *Performance of the FMG–Newton-MG cycle, followed by* 13 *Newton-MG steps, on* $-\Delta u + \gamma u e^u = f$, *with* $\gamma = 10$. *The grid size is* 127×127. *Note that one FMG–Newton-MG cycle reduces the error to the level of discretization error, and that subsequent Newton-MG steps on the fine grid further reduce the residual error to the prescribed tolerance of* 10^{-10}.

Newton-MG cycles on the fine grid. The results are very similar to those for FMG–FAS in Table 6.6. Observe that the norm of the actual error is reduced to the level of discretization error by one FMG–Newton cycle. Subsequent Newton-MG cycles do, however, continue to reduce the discrete L^2 norm of the residual effectively to below the prescribed tolerance.

Both of these methods reduce the error to the level of discretization in one FMG cycle. The flop count indicates that the methods are similar in cost, with the FMG–Newton-MG cycle somewhat less expensive (2.4 Mflops) than the FMG–FAS cycle (3.1 Mflops). However, the individual Newton-MG steps on the fine grid, although cheaper, are not quite as effective as FAS V-cycles for reducing the residual norm. Indeed, if the goal is to reduce the residual norm to 10^{-10}, it is somewhat less expensive (21.2 vs. 26.2 Mflops) to use FAS than Newton-MG. We remind the reader, however, that these flop counts, like those reported earlier, should not be taken too seriously: comparative measures of efficiency depend critically on specific implementation details, computing environment, and problem characteristics. The major conclusion to be reached is that both methods are very efficient and robust.

One of the goals of FMG is to obtain (efficiently) results that are comparable to the accuracy of the discretization. For nonlinear problems, the nested iteration feature of FMG improves the likelihood that initial guesses will lie in the basin of attraction of the chosen iterative method, which accounts for some of its effectiveness. We do not claim that FMG solves all the difficulties that nonlinear problems present. Rather, FMG should be considered a potentially powerful tool that can be used to treat nonlinear problems that arise in practice. ◇◇

So far we have developed FAS in a fairly mechanical way. We showed how the development of multigrid for linear problems can be mimicked to produce a scheme that makes sense for nonlinear equations. However, what we have not done is motivate this development from basic principles in a way that suggests why it works. We attempt to do so now.

Here is the central question in treating nonlinearities by multigrid: When working on the fine grid with the original equation in the form $A^h(\mathbf{v}^h + \mathbf{e}^h) = \mathbf{f}^h$, how can the error for this equation be effectively represented on a coarser grid?

This is analogous to asking: How do you move from the *continuous problem* to the fine grid; that is, how do you effectively discretize the differential equation $A(v + e) = f$ with a known function v? For one possible answer, we turn to the second example above ($-u'' + \gamma uu' = f$) with u replaced by $v + e$:

$$-(v + e)'' + \gamma(v + e)(v + e)' = f.$$

It is important first to think of e and e' as small quantities so that the dominant part of this equation is the term $-v'' + \gamma vv'$, which must be treated carefully. To expose this term in the equation, we expand the products to obtain

$$-(v'' + e'') + \gamma(vv' + v'e + ve' + ee') = f.$$

Since $-v'' + \gamma vv'$ is known, we simply move it to the right side to obtain the *differential residual equation*

$$\gamma(ve' + v'e + ee') = f - (-v'' + \gamma vv'). \tag{6.16}$$

A natural way to discretize this equation begins by evaluating the right side,

$$r = f - (-v'' + \gamma vv') = f - A(v),$$

at the grid points; this produces a vector $r_j^h = r(x_j)$. This way of transferring r from the differential setting to Ω^h is analogous to transferring the fine-grid residual $\mathbf{r}^h$ to Ω^{2h} by injection: $r_j^{2h} = r_{2j}^h$.

For the left side of residual equation (6.16), we need a sensible scheme for evaluating the coefficients v and v' on Ω^h. For v, the natural choice is again to evaluate it at the grid points: $v_j^h = v(x_j)$. For v', we can use a central difference approximation at the grid points:

$$(v')_j^h \approx \frac{v(x_{j+1}) - v(x_{j-1})}{2h}.$$

We are now left with the task of representing e and e' on Ω^h. This is naturally done using the expressions e_j^h and

$$(e')_j^h \approx \frac{e_{j+1}^h - e_{j-1}^h}{2h},$$

respectively.

Putting all of this together gives us a fine-grid approximation to the differential residual equation. We write it at grid point x_{2j} of the fine grid as follows:

$$\gamma v_{2j}^h \left(\frac{e_{2j+1}^h - e_{2j-1}^h}{2h}\right) + \gamma e_{2j}^h \left(\frac{v_{2j+1}^h - v_{2j-1}^h}{2h}\right) + \gamma e_{2j}^h \left(\frac{e_{2j+1}^h - e_{2j-1}^h}{2h}\right)$$
$$= \underbrace{f_{2j}^h - A_{2j}^h(v^h)}_{r_{2j}^h}.$$

We have shown how to move from the continuum to the fine grid in a natural way. How do we make the analogous move from the fine grid to the coarse grid? Terms evaluated at x_{2j} can be restricted directly to the coarse grid; for example, $v_{2j}^h = v_j^{2h}$. The difference approximations can be written as analogous centered differences with respect to the coarse-grid points. Making these replacements, we have the coarse-grid approximation to the fine-grid residual equation:

$$\gamma v_j^{2h} \left(\frac{e_{j+1}^{2h} - e_{j-1}^{2h}}{4h} \right) + \gamma e_j^{2h} \left(\frac{v_{j+1}^{2h} - v_{j-1}^{2h}}{4h} \right) + \gamma e_j^{2h} \left(\frac{e_{j+1}^{2h} - e_{j-1}^{2h}}{4h} \right) = (I_h^{2h} r^h)_j.$$

This is precisely the FAS equation that we derived in the example above.

We just developed FAS as a way to go from the fine to the coarse grid and showed that it can be viewed as a natural extension of the discretization process that goes from the continuum to the fine grid. This exposes a general multigrid coarsening principle: *What is good for discretization is probably good for coarsening.* There are notable exceptions to this principle, but the discretization process is always a good place to start for guidance in constructing the coarse-grid correction scheme.

Along these lines, notice that, in moving from the continuum to the fine grid, we could have computed the approximation to v' by reversing the order of the steps, that is, by *first* computing the derivative v', *then* evaluating it at the grid points: $(v')_i^h \approx v'(x_i)$. When moving from the fine grid to the coarse grid, this choice means we use the fine-grid difference approximation as a coefficient in the coarse-grid equation. Its coarse-grid correction formula is

$$\gamma v_j^{2h} \left(\frac{e_{j+1}^{2h} - e_{j-1}^{2h}}{4h} \right) + \gamma e_j^{2h} \left(\frac{v_{2j+1}^h - v_{2j-1}^h}{2h} \right) + \gamma e_j^{2h} \left(\frac{e_{j+1}^{2h} - e_{j-1}^{2h}}{4h} \right) = (I_h^{2h} r^h)_j.$$

Note that the only difference over FAS is in the coefficient of the second term.

This alternative illustrates that FAS is not necessarily the most direct way to treat the nonlinearity. In many cases, there are coarsening strategies that are more compatible with the discretization processes. See [15] for examples that use discretizations based on projections, as the finite element method does. On the other hand, FAS is advantageous as a general coarsening approach because of its very convenient form

$$A^{2h}(\mathbf{v}^{2h} + \mathbf{e}^{2h}) - A^{2h}(\mathbf{v}^{2h}) = \mathbf{r}^{2h}.$$

Exercises

1. **Linear problems.** Assume that A^h is actually a linear operator: $A^h(\mathbf{u}^h) = A^h \mathbf{u}^h$. Show that formula (6.5) for the FAS coarse-grid correction reduces to the formula for the two-grid correction scheme given in Chapter 3. Do this first by showing that the coarse-grid equations are the same, then by showing that the corrections to the fine-grid approximations are the same.

2. **Fixed point property.** Assume that relaxation has the fixed point property, namely, that if the exact solution $\mathbf{u}^h$ of equation (6.4) is used as the initial guess $\mathbf{v}^h$, then the result is $\mathbf{u}^h$. Assume also that the coarse-grid equation has a unique solution.

(a) Assume that the coarse-grid equation is solved exactly. Show that FAS also has this fixed point property.

(b) Now show that this fixed point property holds even if the exact solver of the coarse-grid equations is replaced by an appropriate approximate solver. Assume only that the coarse-grid solver itself exhibits this fixed point property.

3. **Residual equation.** You may have wondered why the FAS scheme is based on the residual equation and not the original equation $A^h(\mathbf{u}^h) = \mathbf{f}^h$. To see this more clearly, consider the following coarse-grid approximation to $A^h(\mathbf{v}^h + \mathbf{e}^h) = \mathbf{f}^h$:

$$A^{2h}(\mathbf{v}^{2h} + \mathbf{e}^{2h}) = I_h^{2h}\mathbf{f}^h.$$

If this is used in place of the coarse-grid equation $A^{2h}(\mathbf{v}^{2h}+\mathbf{e}^{2h}) = A^{2h}(\mathbf{v}^{2h})+\mathbf{r}^{2h}$, show that FAS generally loses both fixed point properties described in the previous exercise. Thus, for example, if you start with the exact fine grid solution, you will not get it back after the first iteration. Note also that the solution of this coarse-grid equation does not change from one FAS cycle to the next, so that it cannot properly represent smooth components of the fine-grid error, *which do change.*

4. **Example revisited.** Here we show that problem (6.7) of the first example is subtle in the sense that it has a solution if and only if the f_j satisfy a certain compatibility condition. For this purpose, assume that $f_j \neq 0$ for every j.

(a) Remembering that n is an even positive integer, show that if $\mathbf{u}$ satisfies (6.7), then

$$u_1 = \left(\frac{f_1 f_3 \cdots f_{n-1}}{f_2 f_4 \cdots f_n} \right) u_1.$$

(b) Now show that u_1 must be nonzero and that we therefore have the compatibility condition

$$f_1 f_3 \cdots f_{n-1} = f_2 f_4 \cdots f_n.$$

(c) Assuming that this compatibility condition is satisfied, show that the solution to (6.7) is determined up to an arbitrary multiplicative constant.

5. **Scalar Newton's methods within nonlinear Gauss–Seidel.** Show that in solving the component equations of (6.11) for $u_{i,j}$, scalar Newton's method takes the form given in (6.14).

6. **Formulating methods of solution.** Apply parts (a)–(e) to the following boundary value problems:

$$\begin{array}{ll} \text{(i)} \quad u''(x) + e^u = 0, & \text{(ii)} \quad u''(x) + u^2 = 1, \\ \quad u(0) = u(1) = 0. & \quad u(0) = u(1) = 0. \end{array}$$

(a) Use finite difference to discretize the problem. Write out a typical equation of the resulting system of equations.

(b) Determine if nonlinear Gauss–Seidel can be formulated explicitly. If so, write out a typical update step. If not, formulate a typical Newton step that must be used.

(c) Formulate Newton's method for the system and identify the linear system that needs to be solved at each step.

(d) Formulate the solution by FAS, indicating clearly the systems that must be solved on Ω^h and Ω^{2h} .

Chapter 7

Selected Applications

Our model problem for highlighting basic multigrid concepts has been Poisson's equation in one and two dimensions. To see how multigrid can be applied to other problems, we now consider three variations on our model problem, each, we hope, of fairly general interest: Neumann boundary conditions, anisotropic equations, and variable-coefficient/mesh problems. In each case, we present a typical, but simplified, problem that we treat with a specially tailored multigrid technique. The method we propose is not necessarily the most robust, and we ignore complicating issues for the sake of clarity. The purpose here is to illustrate the application of multigrid to new problems and to gain an entry-level understanding of the broad scope of multigrid techniques.

Neumann Boundary Conditions

Dirichlet boundary conditions specify values of the unknown function on the boundary of the domain. In most cases, these so-called *essential boundary conditions* are easily treated in the discretization and in the multigrid solver by simply eliminating the corresponding unknowns in the equations. On the other hand, Neumann conditions involve *derivatives* of the unknown function. This relatively small change in the boundary conditions can be treated in a variety of ways; it also introduces a few important subtleties.

Consider the following two-point boundary value problem with homogeneous Neumann conditions at both end points (the inhomogeneous case is left for Exercise 8):

$$-u''(x) = f(x), \quad 0 < x < 1, \tag{7.1}$$
$$u'(0) = u'(1) = 0.$$

With n an odd positive integer, denote the points of a uniform grid on the interval $[0,1]$ by $x_j = jh$, for $0 \le j \le n+1$, where $h = \frac{1}{n+1}$. A common way to discretize this problem involves the *ghost points* $x_{-1} = -\frac{1}{h}$ and $x_{n+2} = 1 + \frac{1}{h}$. These points are outside the problem domain and are used only temporarily to produce approximations at the boundary points.

With this extended grid, we can use central differences for the differential equation at the interior points *and* boundary points. We also use the ghost points to form central difference approximations for the boundary conditions. These choices lead to the following set of discrete equations:

$$\frac{-u_{j-1}^h + 2u_j^h - u_{j+1}^h}{h^2} = f_j^h, \qquad 0 \le j \le n+1,$$

$$\frac{u_1^h - u_{-1}^h}{2h} = 0,$$

$$\frac{u_{n+2}^h - u_n^h}{2h} = 0.$$

We could now choose to work with this entire system of equations, which requires careful treatment of the boundary equations in relaxation and coarsening. Instead, we eliminate the discrete boundary conditions. The two boundary condition equations imply that

$$u_{-1}^h = u_1^h \quad \text{and} \quad u_{n+2}^h = u_n^h.$$

Eliminating the ghost points in the $j = 0$ and $j = n+1$ equations yields the following $(n+2) \times (n+2)$ system of algebraic equations:

$$\frac{-u_{j-1}^h + 2u_j^h - u_{j+1}^h}{h^2} = f_j^h, \qquad 1 \le j \le n,$$

$$\frac{2u_0^h - 2u_1^h}{h^2} = f_0^h,$$

$$\frac{-2u_n^h + 2u_{n+1}^h}{h^2} = f_{n+1}^h.$$

As before, we write this system in the matrix form

$$A^h \mathbf{u}^h = \mathbf{f}^h, \tag{7.2}$$

where now

$$A^h = \frac{1}{h^2} \begin{bmatrix} 2 & -2 & & & & \\ -1 & 2 & -1 & & & \\ & & \ddots & \ddots & \ddots & \\ & & & \ddots & \ddots & \ddots \\ & & & -1 & 2 & -1 \\ & & & & -2 & 2 \end{bmatrix}. \tag{7.3}$$

Note that system (7.2) involves the boundary unknowns u_0^h and u_{n+1}^h and that A^h is an $(n+2) \times (n+2)$ nonsymmetric matrix.

The first observation to be made is that two-point boundary value problem (7.1) is not yet well-posed. First, if this problem has a solution u, it cannot be unique: because the system involves only derivatives of u, the function $u + c$ must also be a solution for any constant c (Exercise 1). Second and worse yet, we cannot even be sure that boundary value problem (7.1) has a solution. If a solution of (7.1) exists, then the source term must satisfy

$$\int_0^1 f(x)\, dx = 0 \tag{7.4}$$

(Exercise 2). This integral *compatibility condition* is necessary for a solution to exist: if f does not satisfy it, then there can be no solution. Fortunately, the compatibility condition is also sufficient in general. Showing this fact is a little beyond the scope of this book. It amounts to proving that $-\frac{\partial^2}{\partial x^2}$ is a well-behaved operator on the appropriate space of functions u that have zero mean: $\int_0^1 u(x)\,dx = 0$.

This reasoning allows us to conclude that if f satisfies compatibility condition (7.4), then the following boundary value problem is well-posed:

$$
\begin{aligned}
-u''(x) &= f(x), \quad 0 < x < 1, \\
u'(0) = u'(1) &= 0, \\
\int_0^1 u(x)\,dx &= 0.
\end{aligned}
\tag{7.5}
$$

The last condition says that of all the solutions $u + c$ that satisfy the differential equation and boundary conditions, we choose the one with zero mean.

Note that if u solves (7.5), then $u + c$ solves the same boundary value problem with $\int_0^1 u(x)\,dx = c$. Thus, the solution of (7.2) with mean value c is easily found from the zero-mean solution.

In a similar way, discrete system (7.2) is not well-posed. Let $\mathbf{1}^h$ be the $(n+2)$-vector whose entries are all 1, so the entries of $c\mathbf{1}^h$ are all c. The matrix A^h has a nontrivial null space consisting of the constant vectors $c\mathbf{1}^h$ (Exercise 3). Thus, if $\mathbf{u}^h$ is a specific solution of (7.2), then the general solution is $\mathbf{u}^h + c\mathbf{1}^h$ (Exercise 4).

By the Fundamental Theorem of Linear Algebra, (7.2) is solvable only if $\mathbf{f}^h$ is orthogonal to $\mathcal{N}(A^T)$. It is not difficult to show that the $(n+2)$-vector $(\frac{1}{2}, 1, \ldots, 1, \frac{1}{2})$ forms a basis for $\mathcal{N}(A^T)$. Thus, the system is solvable only if $\mathbf{f}$ is orthogonal to $\mathcal{N}((A^h)^T)$, which implies that

$$
\frac{1}{2} f_0^h + \sum_{j=0}^n f_j^h + \frac{1}{2} f_{n+1}^h = 0.
$$

As in the continuous problem, we have two issues to handle: solvability, which requires that $\mathbf{f}^h$ be orthogonal to the null space of $(A^h)^T$, and nonuniqueness, which means that two solutions differ by a vector in the null space of A^h. A useful observation is that if A is symmetric $(A = A^T)$, then solvability and nonuniqueness can be dealt with together, because the null spaces of A and A^T are identical in that case.

Fortunately, there is an extremely simple way to symmetrize this problem: we just divide the first and last equations of the system by 2. This yields a symmetric system whose matrix is

$$
\hat{A}^h = \frac{1}{h^2}
\begin{bmatrix}
1 & -1 & & & & \\
-1 & 2 & -1 & & & \\
& \cdot & \cdot & \cdot & & \\
& & \cdot & \cdot & \cdot & \\
& & & -1 & 2 & -1 \\
& & & & -1 & 1
\end{bmatrix}.
\tag{7.6}
$$

$\hat{A}^h$ *and its transpose have as a null space multiples of the constant vector* $\mathbf{1}^h$. The right-side vector becomes $\hat{\mathbf{f}}^h = [f_0/2, f_1, \ldots, f_n, f_{n+1}/2]^T$. Solvability is then guaranteed by ensuring that $\hat{\mathbf{f}}^h$ is orthogonal to the constant vector $\mathbf{1}^h$:

$$
\langle \hat{\mathbf{f}}^h, \mathbf{1}^h \rangle = \sum_{j=0}^{n+1} \hat{f}_j^h = 0
\tag{7.7}
$$

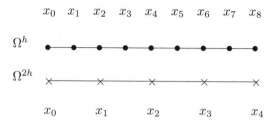

$$x_0 \quad x_1 \quad x_2 \quad x_3 \quad x_4 \quad x_5 \quad x_6 \quad x_7 \quad x_8$$

Figure 7.1: *For the Neumann boundary condition case, the solution is unknown at the end points x_0 and x_{n+1}, where n is an odd integer. The figure shows Ω^h and Ω^{2h} for the case $n = 7$.*

(Exercise 5). This is the discrete analogue of integral compatibility condition (7.4). Hence, the discrete analogue of the well-posed differential system (7.5) is

$$\begin{aligned}
\frac{-u_{j-1}^h + 2u_j^h - u_{j+1}^h}{h^2} &= f_j^h, \qquad 1 \le j \le n, \\
\frac{u_0^h - u_1^h}{h^2} &= \frac{f_0^h}{2}, \\
\frac{-u_n^h + u_{n+1}^h}{h^2} &= \frac{f_{n+1}^h}{2}, \\
\sum_{i=0}^{n+1} u_i^h &= 0,
\end{aligned} \tag{7.8}$$

or simply

$$\hat{A}^h \mathbf{u}^h = \hat{\mathbf{f}}^h, \tag{7.9}$$

$$\langle \mathbf{u}^h, \mathbf{1}^h \rangle = 0. \tag{7.10}$$

As before, of all solutions to the problem, that last condition selects the solution with zero mean. Matrix system (7.9)–(7.10) is well-posed in the sense that it has a unique solution, provided $\hat{\mathbf{f}}^h$ satisfies (7.7).

The ingredients of our multigrid algorithm for solving (7.9)–(7.10) are basically the same as before, except we now must account for the extra equations associated with the unknowns at the boundary points and the zero-mean condition. Figure 7.1 shows a fine grid, Ω^h, with $n + 2 = 9$ points and the corresponding coarse grid Ω^{2h} with $\frac{n+3}{2} = 5$ points.

The approximation $\mathbf{v}^h$ must include the entries v_0^h and v_{n+1}^h, and the coarse-grid correction $\mathbf{v}^{2h}$ must include v_0^{2h} and $v_{(n+1)/2}^{2h}$. Relaxation is now performed at *all* the fine-grid points, including x_0 and x_{n+1}. Relaxation at the end points is based on the second and third equations of (7.8):

$$\begin{aligned}
v_0^h &\leftarrow v_1^h + h^2 \hat{f}_0^h, \\
v_{n+1}^h &\leftarrow v_n^h + h^2 \hat{f}_{n+1}^h.
\end{aligned}$$

To account for zero-mean condition (7.10), we now just add the step

$$\mathbf{v}^h \leftarrow \mathbf{v}^h - \frac{\langle \mathbf{v}^h, \mathbf{1}^h \rangle}{\langle \mathbf{1}^h, \mathbf{1}^h \rangle} \mathbf{1}^h, \tag{7.11}$$

which is just the Gram–Schmidt method applied to orthogonality condition (7.10). (Note that $\langle \mathbf{1}^h, \mathbf{1}^h \rangle = n + 2$.) This global step can be applied before or after relaxation on each level. In principle, this step can wait until the very end of the multigrid algorithm since its primary role is to produce an approximation with the correct average. From a mathematical point of view, it does not really matter if the intermediate approximation $\mathbf{v}^h$ has a nonzero mean. However, from a numerical point of view, it is probably best to apply (7.11) often enough to prevent a large constant from creeping in and swamping accuracy.

The coarsening process is similar to that for the Dirichlet case, but now we must remember to interpolate corrections to the end point approximations. These end points are present on both fine and coarse grids, so the correction process is straightforward. For example, recalling that the terms v_j^{2h} are errors computed on the coarse grid, the correction steps are given by

$$
\begin{aligned}
v_0^h &\leftarrow v_0^h + v_0^{2h}, & (7.12) \\
v_1^h &\leftarrow v_1^h + \frac{v_0^{2h} + v_1^{2h}}{2}.
\end{aligned}
$$

What about restriction of residuals to the coarse grid? In the Dirichlet case, we chose I_h^{2h} to be the transpose of I_{2h}^h scaled by $\frac{1}{2}$:

$$
I_h^{2h} = \frac{1}{2} \left(I_{2h}^h \right)^T .
$$

We will use the same restriction operator here, but we must determine its form at the boundary points. Correction step (7.12) shows that interpolation relates v_0^{2h} to v_0^h using the weight 1 and relates v_0^{2h} to v_1^h using the weight $\frac{1}{2}$. Thus, reversing these relationships (which is what the transpose does) and scaling them by $\frac{1}{2}$, we arrive at the restriction step (Exercise 6)

$$
\hat{f}_0^{2h} \leftarrow \frac{1}{2} \hat{f}_0^h + \frac{1}{4} \hat{f}_1^h .
$$

The coarse-grid matrix we get from the Galerkin condition, $\hat{A}^{2h} = I_h^{2h} \hat{A}^h I_{2h}^h$, is just the coarse-grid version of $\hat{A}^h$ defined in (7.3): this follows from Table 5.1 for the interior points, and from Table 7.1 for the boundary point $x = 0$ ($x = 1$ is analogous). One consequence of this relationship is that the multigrid scheme preserves the variational properties we used in the Dirichlet case.

Another subtlety that we need to address is the discrete compatibility condition. If we are given a general source vector $\hat{\mathbf{f}}^h$, we can be sure that the fine-grid problem has a solution simply by testing the zero-mean condition (7.7). But what do we do if this test fails? We could simply stop because this problem has no solution. However, we might find it advantageous to solve the nearby problem created by making the replacement

$$
\hat{\mathbf{f}}^h \leftarrow \hat{\mathbf{f}}^h - \frac{\langle \hat{\mathbf{f}}^h, \mathbf{1}^h \rangle}{\langle \mathbf{1}^h, \mathbf{1}^h \rangle} \mathbf{1}^h . \tag{7.13}
$$

But what about the coarse-grid problem? The answer is that the coarse-grid equation will be solvable, at least in theory, whenever the fine-grid equation is solvable (Exercise 7). However, to be sure that numerical round-off does not perturb the

	0		1
$\hat{e}_0^{2h}$	1		0
$I_{2h}^h \hat{e}_0^{2h}$	1	$\frac{1}{2}$	0
$\hat{A}^h I_{2h}^h \hat{e}_0^{2h}$	$\frac{1}{2h^2}$	0	$-\frac{1}{2h^2}$
$I_h^{2h} \hat{A}^h I_{2h}^h \hat{e}_0^{2h}$	$\frac{1}{4h^2}$		$-\frac{1}{4h^2}$

Table 7.1: *Calculation of the first row of* $\hat{A}^{2h} = I_h^{2h} \hat{A}^h I_{2h}^h$ *at the boundary* $x = 0$.

solvability too much, it is probably best to incorporate a Gram–Schmidt step analogous to (7.13) on the coarse grids:

$$\hat{\mathbf{f}}^{2h} \leftarrow \hat{\mathbf{f}}^{2h} - \frac{\langle \hat{\mathbf{f}}^{2h}, \mathbf{1}^{2h} \rangle}{\langle \mathbf{1}^{2h}, \mathbf{1}^{2h} \rangle} \mathbf{1}^{2h} \, .$$

Note that $\langle \mathbf{1}^{2h}, \mathbf{1}^{2h} \rangle = \frac{n+3}{2}$.

Numerical example. Consider the two-point boundary value problem

$$\begin{aligned} -u''(x) &= 2x - 1, \qquad 0 < x < 1, \\ u'(0) = u'(1) &= 0. \end{aligned}$$

By integrating the differential equation twice and applying the boundary conditions, we find that the function $u(x) = \frac{x^2}{2} - \frac{x^3}{3} + c$ solves the problem for any constant c. The zero-mean solution corresponds to $c = -\frac{1}{12}$, and we use this function as our exact solution in the numerical experiments. Solutions are approximated on a succession of grids to illustrate the effectiveness of the algorithm as n increases.

Symmetrization of the problem produces the source vector $\hat{\mathbf{f}}$, where $\hat{f}_i = f_i$ for $1 \le i \le n$,

$$\hat{f}_0 = \frac{f_0}{2} = -\frac{1}{2}, \quad \text{and} \quad \hat{f}_{n+1} = \frac{f_{n+1}}{2} = \frac{1}{2}.$$

It can be checked that $\hat{\mathbf{f}}$ satisfies discrete compatibility condition (7.7). The restriction, interpolation, and coarse-grid operators described above are used with V(2,1)-cycles based on Gauss–Seidel relaxation. For this problem, which is computed in double precision, we observed that it was enough to apply the Gram–Schmidt process once at the end of each V-cycle.

Table 7.2 shows the results of this experiment. Each test is stopped when the discrete L^2 norm of the residual drops below 10^{-10}. Listed are the grid sizes, the final residual norm, the average convergence factor, the final error norm (the difference between the computed solution and the sampled version of the continuous solution), and the number of V-cycles required to reach the desired tolerance.

| Grid size | | Average | | Number |
n	$\|r^h\|_h$	conv. factor	$\|e\|_h$	of cycles
31	6.3e−11	0.079	9.7e−05	9
63	1.9e−11	0.089	2.4e−05	10
127	2.6e−11	0.093	5.9e−06	10
255	3.7e−11	0.096	1.5e−06	10
511	5.7e−11	0.100	3.7e−07	10
1027	8.6e−11	0.104	9.2e−08	10
2047	2.1e−11	0.112	2.3e−08	10
4095	5.2e−11	0.122	5.7e−09	11

Table 7.2: *Numerical results for* $-u''(x) = f(x), u'(0) = u'(1) = 0$. *Shown are discrete L^2 norms of the final residual and the final error, average convergence factors, and number of* V(2,1)-*cycles required to reduce the residual norm to less than* 10^{-10}.

It is evident from the table that the method is very robust on this problem in that the speed of convergence is essentially independent of problem size. Furthermore, with the strict tolerance on the residual norm used, the discrete problems are solved well below the level of discretization error. The final errors represent the discretization errors themselves: they decrease by almost exactly one-fourth with each doubling of n, which is evidence of $O(h^2)$ behavior.

An interesting subtlety of this problem is that we could have chosen to use *one-sided differences* for approximating the boundary derivatives:

$$u'(0) \approx \frac{u_1 - u_0}{h} \quad \text{and} \quad u'(1) \approx \frac{u_{n+2} - u_{n+1}}{h}.$$

This choice would have resulted in exactly the same operator matrix, $\hat{A}^h$, that we obtained by symmetrizing the discretization based on central differences. However, the source vector would be the original vector, $\mathbf{f}$, rather than the vector $\hat{\mathbf{f}}$. Using this system, and leaving the remainder of the algorithm unchanged, we found experimentally that the speed of convergence and residual errors were approximately as we see in Table 7.2, as expected. However, the final errors (that is, the discretization errors) were much larger and did not decrease nearly as fast with increasing n, because one-sided differences are not as accurate as central difference (($O(h)$ compared to $O(h^2)$). ◇◇

Anisotropic Problems

The problems treated thus far have been limited to matrices with constant nonzero off-diagonal entries, namely, $-\frac{1}{h^2}$. Such matrices naturally arise in treating Poisson's equation on uniform grids. Our first departure from this situation is to consider two-dimensional problems in which the matrices have two different constants appearing in the off-diagonal terms. Such matrices arise when

- the differential equation has constant, but different, coefficients for the derivatives in the coordinate directions, or

- when the discretization has constant but different mesh sizes in each coordinate direction.

The model problems for these two cases are, respectively, the differential equation

$$-u_{xx} - \epsilon u_{yy} = f \qquad (7.14)$$

discretized on a uniform grid of mesh size h and Poisson's equation ($\epsilon = 1$) discretized on a grid with constant mesh size $h_x = h$ in the x-direction and constant mesh size $h_y = \frac{h}{\sqrt{\epsilon}}$ in the y-direction. To be very distinct from the isotropic case $\epsilon = 1$, we assume that $0 < \epsilon << 1$.

It is interesting that these two *anisotropic* model problems lead to the same five-point stencil (Exercise 9):

$$A^h = \frac{1}{h^2} \begin{pmatrix} & -\epsilon & \\ -1 & 2 + 2\epsilon & -1 \\ & -\epsilon & \end{pmatrix}. \qquad (7.15)$$

This relationship between problems with variable coefficients and problems with variable mesh sizes is important; it means that we can think of either example as we develop a method to treat anisotropic problems. Note that the *weak connection* in the vertical direction in these examples arises, respectively, from a small coefficient of the y-derivative term or a large grid spacing in the y-direction.

This departure from the model (isotropic) problem unfortunately leads to trouble with the standard multigrid approach: multigrid convergence factors degrade as ϵ tends to zero, and even for $\epsilon \approx 0.1$, poor performance can be expected.

To understand why multigrid deteriorates, consider (7.14) in the limiting case $\epsilon = 0$. The matrix then becomes

$$A^h = \frac{1}{h^2} \begin{pmatrix} & 0 & \\ -1 & 2 & -1 \\ & 0 & \end{pmatrix},$$

which means that the discrete problem becomes a collection of one-dimensional Poisson equations in the x-direction, with no connections in the y-direction. Gauss–Seidel or damped Jacobi point relaxation will smooth in the x-direction because it does so in one dimension. However, the lack of connections in the y-direction means that errors on one x-line have nothing to do with errors on any other: errors in the y-direction will generally have a random pattern, far from the smoothness needed for coarsening to work well.

We can gain further insight into the problems that arise with anisotropy if we look at the eigenvalues of the iteration matrix. Recalling Exercise 12 of Chapter 4, if the weighted Jacobi method with parameter ω is applied to the model Poisson equation in two dimensions on an $n \times n$ grid, the eigenvalues of the iteration matrix are given by

$$\lambda_{k,\ell} = 1 - \omega \left(\sin^2 \left(\frac{k\pi}{2n} \right) + \sin^2 \left(\frac{\ell\pi}{2n} \right) \right), \qquad 1 \le k, \ell \le n.$$

The wavenumbers (frequencies) k and ℓ correspond to the x- and y-directions, respectively. The same analysis (Exercise 10) applied to the problem (7.14) reveals that the eigenvalues of the Jacobi iteration matrix are

$$\lambda_{k,\ell} = 1 - \frac{2\omega}{1+\epsilon} \left(\sin^2\left(\frac{k\pi}{2n}\right) + \epsilon \sin^2\left(\frac{\ell\pi}{2n}\right) \right), \quad 1 \le k, \ell \le n.$$

Notice that for small ϵ, the contributions from the wavenumbers in the y-direction (the direction of weak coupling) are insignificant. Thus, there is little variation in the eigenvalues with respect to the y-wavenumbers. The variation in eigenvalues with respect to the x-wavenumbers is what we expect in the one-dimensional problem.

The eigenvalue picture is best given in the two plots of Fig. 7.2, with $\epsilon = 0.05, n = 16$, and $\omega = \frac{2}{3}$. The upper plot shows the variation in eigenvalues with respect to ℓ (the y-wavenumber) on lines of constant k; notice that there is little variation in the eigenvalues along a single line in this direction. The lower plot shows the variation in eigenvalues with respect to k (the x-wavenumber) on lines of constant ℓ. This time we see individual curves that look much like the eigenvalue curves for the one-dimensional problem. In addition, the curves are tightly bunched, meaning that the convergence is much the same along any horizontal line of the grid. It should be mentioned that local mode analysis, as discussed in Chapter 4, could have been used to reach many of these same conclusions.

These observations suggest two possible strategies for dealing with the anisotropy:

- Because we expect typical multigrid convergence for the one-dimensional problems along x-lines, we should "do multigrid" and coarsen the grid in the x-direction, but not in the y-direction.

- Because the equations are strongly coupled in the x-direction, it will prove advantageous to solve for entire lines of unknowns in the x-direction all at once; this is the goal of *line* or *block relaxation*.

We now investigate these two strategies in more detail.

Semicoarsening/point relaxation

Because relaxation smooths in the x-direction, it makes sense to coarsen horizontally by eliminating every other vertical (y-) line. However, because point relaxation does not smooth in the y-direction, we should *not* coarsen vertically; all horizontal lines should be retained on the coarse grid. This semicoarsening process is depicted in Fig. 7.3. This means that when we write Ω^{2h}, we really mean the coarse grid that has the original grid spacing in the y-direction and twice the original grid spacing in the x-direction.

Interpolation can be done in a one-dimensional way along each horizontal line, giving coarse-grid correction equations of the form

$$v_{2i,j}^h \leftarrow v_{2i,j}^h + v_{i,j}^{2h}, \quad v_{2i+1,j}^h \leftarrow v_{2i+1,j}^h + \frac{v_{i,j}^{2h} + v_{i+1,j}^{2h}}{2}.$$

Semi-coarsening is not as "fast" as *full* coarsening: going from the fine to the coarse grid, the number of points is reduced by a factor of about two with semicoarsening, as opposed to the usual factor of about four. This means that W-cycles lose $O(n)$ complexity; however, V-cycle complexity remains $O(n)$ (Exercise 12).

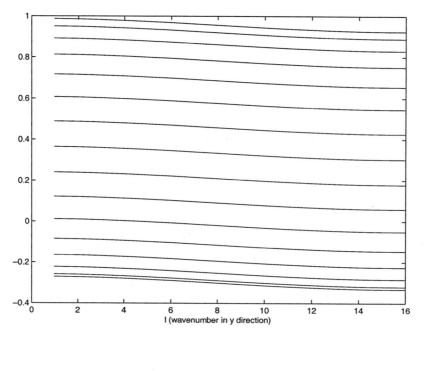

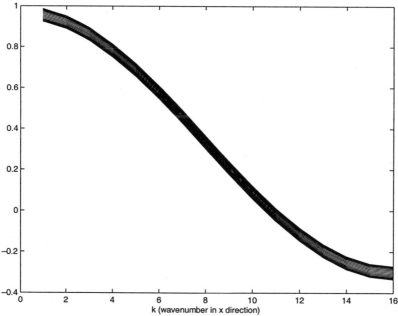

Figure 7.2: *The upper plot shows the eigenvalues of the Jacobi iteration matrix, with $\omega = \frac{2}{3}$, along lines of constant k (wavenumber in the x-direction). The upper curve corresponds to $k = 1$ and the lower curve to $k = n = 16$. The lower plot shows the same eigenvalues along lines of constant ℓ (wavenumber in the y-direction). The small parameter in (7.14) is $\epsilon = 0.05$.*

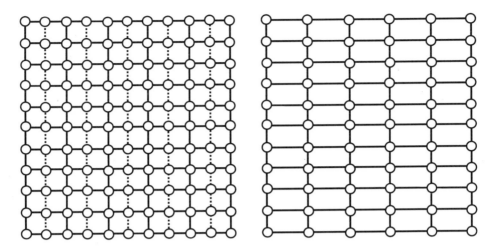

Figure 7.3: *The original fine grid (left, solid and dashed lines) is semicoarsened in the x-direction by deleting the dashed lines (right).*

Line relaxation/full coarsening

Line (or block) relaxation can be developed by writing a system of linear equations in block form. In the case of (7.14), if we order the unknowns along lines of constant y (because strong coupling is in the x-direction), the matrix A^h can be written in block form as

$$A^h = \begin{bmatrix} D & -cI & & & \\ -cI & D & -cI & & \\ \cdot & -cI & D & -cI & \\ & \cdot & \cdot & \cdot & -cI \\ & & & -cI & D \end{bmatrix}, \quad (7.16)$$

where $c = \frac{\epsilon}{h^2}$ and I is the identity matrix. For this model problem, the diagonal blocks D are tridiagonal and identical. Each block is associated with an individual horizontal grid line and has the stencil $\frac{1}{h^2}(-1 \quad 2 + 2\epsilon \quad -1)$.

One sweep of the *block Jacobi method* consists of solving a tridiagonal system for each line of constant y. The jth system of the sweep has the form

$$D\mathbf{v}_j^h = \mathbf{g}_j^h,$$

where $\mathbf{v}_j^h$ is the jth subvector of $\mathbf{v}^h$ with entries $\left(\mathbf{v}_j^h\right)_i = v_{i,j}^h$, and $\mathbf{g}_j^h$ is the jth right-side vector with entries

$$\left(\mathbf{g}_j^h\right)_i = f_{i,j}^h + \frac{\epsilon}{h^2}\left(v_{i,j-1}^h + v_{i,j+1}^h\right).$$

Because D is tridiagonal, these systems can be solved efficiently with some form of Gaussian elimination. The operation count for relaxation remains $O(n)$, and so does the cost of either V- *or* W-cycle solvers. A weighted block Jacobi method results from averaging the current approximation and a full Jacobi update using a damping parameter ω.

To see exactly why line relaxation in the direction of strong coupling is effective, we need to look at the convergence properties of the iteration matrix. A brisk

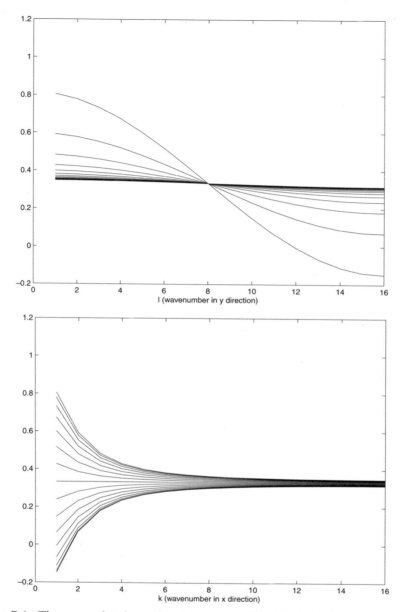

Figure 7.4: *The upper plot shows the eigenvalues of the block Jacobi iteration matrix, with $\omega = \frac{2}{3}$, along lines of constant k (wavenumber in the x-direction). The upper curve corresponds to $k = 1$ and the flattest curve to $k = n = 16$. The lower plot shows the same eigenvalues along lines of constant ℓ (wavenumber in the y-direction). The upper curve corresponds to $\ell = 1$ and the lower curve corresponds to $\ell = n = 16$. The small parameter in (7.14) is $\epsilon = 0.05$.*

calculation (Exercise 11) confirms that the eigenvalues of the *block* Jacobi iteration matrix (with damping parameter ω) are given by

$$\lambda_{k,\ell} = 1 - \frac{2\omega}{2\sin^2(\frac{k\pi}{2n}) + \epsilon} \left(\sin^2\left(\frac{k\pi}{2n}\right) + \epsilon \sin^2\left(\frac{\ell\pi}{2n}\right) \right), \quad 1 \le k, \ell \le n.$$

As before, k and ℓ are the wavenumbers in the x- and y-directions, respectively. The eigenvalues for the block Jacobi method (with $\omega = \frac{2}{3}$ and $\epsilon = 0.05$) are displayed in Fig. 7.4, which is analogous to Fig. 7.2 for the point Jacobi method. The top figure shows the eigenvalues along lines of constant k; the lower figure shows the same eigenvalues along lines of constant ℓ. The most striking difference between the eigenvalue plots for the two Jacobi methods is the reduction in the size of the eigenvalues: the maximum magnitude of the eigenvalues of the point method is 0.98, while the maximum magnitude of the eigenvalues of the block method is 0.81. This reduction in the size of the eigenvalues results in improved convergence. This improvement is typical of block methods that collect additional coefficients in the diagonal blocks. However, for anisotropic problems, the improvement occurs only if the coefficients in the direction of strong coupling appear in the diagonal blocks.

Both of these strategies are effective for the anisotropic models. Either one can be used when there is a weak connection in a coordinate direction that is known beforehand. Choosing between the two often depends on the nature of the application and the computer architecture involved. However, problems in which the weak direction is unknown or changing within the domain may not be efficiently handled by either method. For more robustness, we could use point relaxation and alternate semicoarsening between the x- and y-directions. However, the increased number of coarse grids that this approach requires is too awkward for most applications. We could also choose to apply full coarsening and alternating line relaxation that switches between x- and y-line relaxation. This approach can also be a bit awkward for some applications, particularly as it extends to three dimensions. Another choice is to apply *both* semi-coarsening and line relaxation.

Semicoarsening/line relaxation

Suppose we want to develop one method that can handle either of the following stencils:

$$A_1^h = \frac{1}{h^2} \begin{pmatrix} & -\epsilon & \\ -1 & 2 + 2\epsilon & -1 \\ & -\epsilon & \end{pmatrix} \text{ or } A_2^h = \frac{1}{h^2} \begin{pmatrix} & -1 & \\ -\epsilon & 2 + 2\epsilon & -\epsilon \\ & -1 & \end{pmatrix}.$$

Semi-coarsening in the x-direction could be used to handle A_1^h and y-line relaxation to handle A_2^h. If we do both, the problem may be viewed as a stack of pencils in the y-direction as shown in Fig. 7.5. Line relaxation in the y-direction is used to solve the problem associated with each pencil, and coarsening is done simply by deleting every other pencil. There is no assumption here about the direction of weak connections; in fact, this approach applies well to the isotropic case. Again, we lose the $O(n)$ complexity of W-cycles, but V-cycle complexity remains $O(n)$ (Exercise 13).

Numerical example. Many of the above ideas can be illustrated with a fairly simple example. Consider equation (7.14),

$$-u_{xx} - \epsilon u_{yy} = f,$$

with homogeneous Dirichlet boundary conditions. The discretized problem has the stencil given by (7.15). We apply three different schemes to this problem for

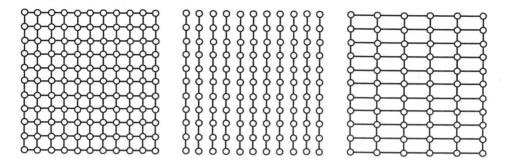

Figure 7.5: *It is possible to combine line relaxation and semicoarsening. The original grid (left) is viewed as a collection of pencils for line relaxation in the y-direction (center) and semi-coarsened in the x-direction (right).*

selected values of ϵ, ranging from large values, so that coupling is dominant in the y-direction, to very small values, so that coupling is dominant in the x-direction. Our example uses the right-side function

$$f(x,y) = 2(y - y^2) + 2\epsilon(x - x^2),$$

which produces the exact solution

$$u(x,y) = (x - x^2)(y - y^2).$$

The first approach consists of a standard V(2,1)-cycle, based on Gauss–Seidel relaxation, full coarsening, full weighting, and linear interpolation. We do not expect this method to be effective, except when $\epsilon \approx 1$. A simple demonstration pinpoints the difficulty we expect to encounter in this experiment.

Letting $\epsilon = 0.001$ and using a 16×16 grid with $n = 16$ points, we begin the iteration with a random initial guess. Figure 7.6 shows a surface plot of the error after 50 sweeps of Gauss–Seidel. We also show the error along a line of constant y (middle figure) and a line of constant x (bottom figure). Relaxation apparently smooths the error nicely in the x-direction, the direction of strong coupling, but leaves highly oscillatory error in the y-direction. While it is easy to envision approximating the smooth error curve on a coarser grid, we could not hope to do so with the oscillatory error. Therefore, we must keep all the points in the y-direction if we wish to represent the error accurately in that direction.

These observations lead to the second approach, which is semicoarsening. Because smoothing in the y-direction with point Gauss–Seidel is ineffective when ϵ is small, we coarsen *only* in the x-direction. This means that the coarse and fine grids have the same number of points in the y-direction. Full weighting and linear interpolation are used, but in a one-dimensional way that involves neighbors in the x-direction only. Semicoarsening means that we do not use neighbors in the y-direction for either restriction or interpolation.

It is important to notice that once the semicoarsening has been performed, the discrete operator must be altered to fit the new geometry. That is, because the grid spacing is no longer the same in the two directions, the residual and relaxation

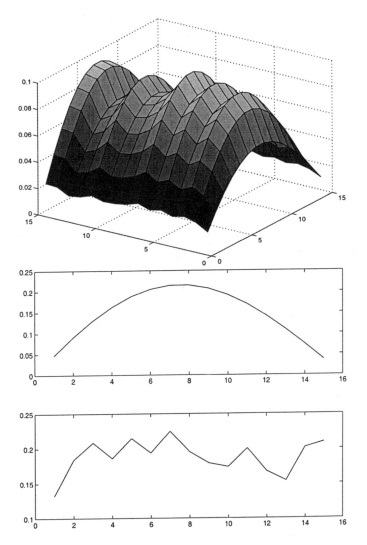

Figure 7.6: *Top: Error after 50 sweeps of pointwise Gauss–Seidel on the equation $-u_{xx} - \epsilon u_{yy} = 0$, beginning with a random initial guess. Middle: Error in the strongly coupled x-direction along a line of constant y. Bottom: Error along a line of constant x, in the weakly coupled y-direction.*

operators must use a stencil like

$$
A = \begin{pmatrix}
 & -\dfrac{\epsilon}{h_y^2} & \\[2ex]
-\dfrac{1}{h_x^2} & \left(\dfrac{2}{h_x^2} + \dfrac{2\epsilon}{h_y^2}\right) & -\dfrac{1}{h_x^2} \\[2ex]
 & -\dfrac{\epsilon}{h_y^2} &
\end{pmatrix},
$$

Scheme	ϵ								
	1000	100	10	1	0.1	0.01	0.001	0.0001	0
V(2,1)-cycles	0.95	0.94	0.58	0.13	0.58	0.90	0.95	0.95	0.95
Semi-C	0.99	0.99	0.98	0.93	0.71	0.28	0.07	0.07	0.07
Semi-C/line relax	0.04	0.08	0.08	0.08	0.07	0.07	0.07	0.08	0.08

Table 7.3: *Convergence factors for the three different schemes for various values of ϵ. The V(2,1) scheme is just standard multigrid, with Gauss–Seidel relaxation, linear interpolation, and full weighting. The semi-C scheme uses semicoarsening in the strongly coupled x-direction. The semi-C/line relax scheme combines semicoarsening in the x-direction with line relaxation in the y-direction.*

where h_x and h_y are the grid spacings in the x- and y-directions, respectively. Our earlier discussion suggests that this scheme will work well when the coupling is dominant in the x-direction (small ϵ), but that it may work poorly otherwise.

The third approach we consider is appropriate when the size of ϵ is unknown, so that the anisotropy could be in either coordinate direction. Here we use semicoarsening in the x-direction and line relaxation in the y-direction. Again, we use full weighting and linear interpolation with neighbors in the x-direction only. Semicoarsening should take care of strong coupling in the x-direction, while line relaxation should handle strong coupling in the y-direction.

The asymptotic convergence factors for the three approaches and for various choices of ϵ are displayed in Table 7.3. The standard V(2,1)-cycle approach works well only when ϵ is very near 1. It converges poorly for values of ϵ that differ from 1 by an order of magnitude or more. The semicoarsening approach performs extremely well when ϵ is very small, but its convergence deteriorates even when $\epsilon = 0.01$ and 0.1. There is a subtle lesson here. Semicoarsening is effective provided $\frac{1}{h_x^2}$ remains larger than $\frac{\epsilon}{h_y^2}$ as h_x is doubled. This means that the coarse grids see much the same anisotropy as the fine grid. However, if $\frac{1}{h_x^2}$ and $\frac{\epsilon}{h_y^2}$ become comparable in size during the x-coarsening process, then convergence rates will decrease. This effect can be observed in Table 7.3. With $n = 16$ and $\epsilon = 0.1$, coarsening in the x-direction quickly produces coupling in the y-direction. Rather than continue to semicoarsen in the x-direction, it is best to return to standard coarsening in this situation.

Finally, the combination of semicoarsening with line relaxation is extremely robust, giving excellent convergence results regardless of the degree of anisotropy.

◇◇

Variable-Mesh Problems

Nonuniform grids are commonly used in practice to accommodate irregularities in the problem domain or multiple scales in the emerging solution. To see how multigrid might handle such problems, consider the one-dimensional Poisson equation

$$-u''(x) \;=\; f(x), \quad 0 < x < 1, \qquad (7.17)$$
$$u(0) = u(1) \;=\; 0.$$

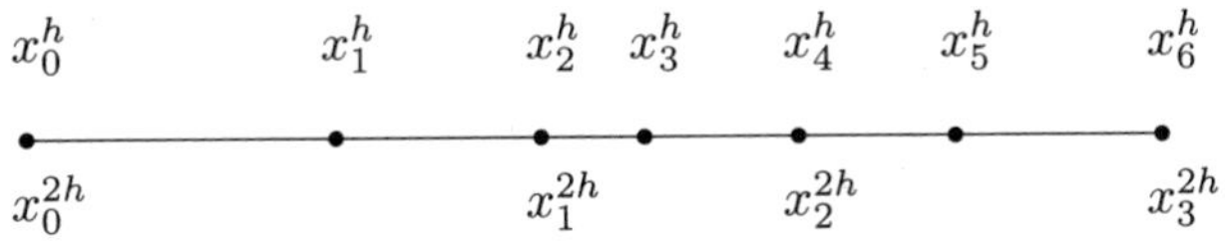

$$x_0^h \qquad x_1^h \qquad x_2^h \; x_3^h \qquad x_4^h \qquad x_5^h \qquad x_6^h$$

$$x_0^{2h} \qquad\qquad x_1^{2h} \qquad x_2^{2h} \qquad\qquad x_3^{2h}$$

Figure 7.7: *With a variable mesh, the coarse grid is defined as the even-numbered grid points of the fine grid.*

With n a positive even integer, define a nonuniform grid on $[0,1]$ by the points x_j^h, $0 \le j \le n$, and mesh sizes $h_{j+\frac{1}{2}} \equiv x_{j+1}^h - x_j^h$, $0 \le j \le n-1$. Using second-order finite differences, we can discretize (7.17) on this grid. The result is (Exercise 14)

$$-\alpha_j^h u_{j-1}^h + \left(\alpha_j^h + \beta_j^h\right) u_j^h - \beta_j^h u_{j+1}^h = f_j^h, \quad 1 \le j \le n-1, \tag{7.18}$$

where the coefficients are given by

$$\alpha_j^h = \frac{2}{h_{j-\frac{1}{2}}\left(h_{j-\frac{1}{2}} + h_{j+\frac{1}{2}}\right)}, \quad \beta_j^h = \frac{2}{h_{j+\frac{1}{2}}\left(h_{j-\frac{1}{2}} + h_{j+\frac{1}{2}}\right)}, \quad 1 \le j \le n-1. \tag{7.19}$$

Note that the operator in (7.18) reduces to the standard stencil $\frac{1}{h^2}\begin{pmatrix} -1 & 2 & -1 \end{pmatrix}$ in the uniform grid case, $h_{i+\frac{1}{2}} = h$.

We can solve this discrete system by modifying the standard multigrid scheme to accommodate variable-mesh sizes. The operator I_{2h}^h is again based on linear interpolation, but it must now account for the nonuniformity of the grid. Choosing the coarse grid (Fig. 7.7) to consist of every other fine-grid point, $x_j^{2h} \equiv x_{2j}^h$, $0 \le j \le \frac{n}{2}$, the interpolation process, $\mathbf{v}^h = I_{2h}^h \mathbf{v}^{2h}$, is defined by

$$v_{2j}^h = v_j^{2h}, \tag{7.20}$$

$$v_{2j+1}^h = \frac{h_{2j+\frac{3}{2}} v_j^{2h} + h_{2j+\frac{1}{2}} v_{j+1}^{2h}}{h_{2j+\frac{1}{2}} + h_{2j+\frac{3}{2}}}, \quad 1 \le j \le \frac{n}{2} - 1.$$

This formula comes from constructing a linear function in the interval $[x_j^{2h}, x_{j+1}^{2h}]$ that interpolates $\mathbf{v}^{2h}$ at the end points (Exercise 15).

Another way to derive (7.20) is to apply *operator interpolation*. The basic idea is to assume that relaxation does not significantly change the error, so that the residual must be almost zero. Using (7.18) at x_{2j+1}^h, we have

$$-\alpha_{2j+1}^h e_{2j}^h + \left(\alpha_{2j+1}^h + \beta_{2j+1}^h\right) e_{2j+1}^h - \beta_{2j+1}^h e_{2j+2}^h \approx 0. \tag{7.21}$$

Solving for e_{2j+1}^h yields

$$e_{2j+1}^h \approx \frac{\alpha_{2j+1}^h e_{2j}^h + \beta_{2j+1}^h e_{2j+2}^h}{\alpha_{2j+1}^h + \beta_{2j+1}^h}.$$

This suggests that we define interpolation by the formula

$$v_{2j}^h = v_j^{2h}, \tag{7.22}$$

$$v_{2j+1}^h = \frac{\alpha_{2j+1}^h v_j^{2h} + \beta_{2j+1}^h v_{j+1}^{2h}}{\alpha_{2j+1}^h + \beta_{2j+1}^h}, \quad 1 \le j \le \frac{n}{2} - 1.$$

It is important that linear interpolation (7.20) and operator interpolation (7.22) be identical for our one-dimensional model problem (Exercise 16). Note that both (7.20) and (7.22) reduce to a simple average for the uniform grid case.

Full weighting can now be defined by the variational property

$$I_h^{2h} \equiv \frac{1}{2} \left(I_{2h}^h\right)^T$$

and the coarse-grid operator can be constructed from the Galerkin principle

$$A^{2h} = I_h^{2h} A^h I_{2h}^h.$$

This definition of A^{2h} yields a stencil that is similar, but not identical, to the stencil obtained by using (7.18)–(7.19) on the coarse grid. However, the Galerkin principle does hold if this system is rescaled by $(h_{j-\frac{1}{2}} + h_{j+\frac{1}{2}})/2$ (Exercise 17).

Variable-mesh problems in two dimensions can be handled in a similar way, but the first issue in this case is the discretization itself. Forming an accurate discrete approximation can be tricky on nonuniform grids in two and higher dimensions. Moreover, if the grid is *logically rectangular* (it can be indexed simply by i, j), then coarsening is generally straightforward; but unstructured grids can make coarsening seem almost impossible. Instead of treating such cases here, we defer this issue to our discussion of the algebraic multigrid algorithm (Chapter 8), which was designed just for this purpose.

Variable-Coefficient Problems

Another common practical problem is the solution of variable coefficient differential equations. Again we consider a simple case to see how multigrid might handle such problems. Our model problem is the scalar elliptic equation

$$\begin{aligned}
-(a(x)u')'(x) &= f(x), \quad 0 < x < 1, \\
u(0) = u(1) &= 0,
\end{aligned} \tag{7.23}$$

where $a(x)$ is a positive function on the interval $(0, 1)$. A *conservative* or *self-adjoint* method for discretizing (7.23) is developed as follows.

Letting n be a positive even integer, consider the *uniform* grid defined by the points $x_j^h = \frac{j}{n}$, $0 \le j \le n$. As usual, suppose that the vector $\mathbf{v}^h$ approximates $\mathbf{u}$ at these grid points, but that values of the coefficient a are taken at the cell centers $x_{j+\frac{1}{2}}^h = \frac{j+\frac{1}{2}}{n}$; thus, we have $a_{j+\frac{1}{2}}^h = a(x_{j+\frac{1}{2}}^h)$ for $0 \le j \le n-1$. The discrete system that results (Exercise 18) is

$$\frac{1}{h^2}\left(-a_{j-\frac{1}{2}}^h u_{j-1}^h + (a_{j-\frac{1}{2}}^h + a_{j+\frac{1}{2}}^h)u_j^h - a_{j+\frac{1}{2}}^h u_{j+1}^h\right) = f_j^h, \quad 1 \le j \le n-1,$$

$$u_0^h = u_n^h = 0. \tag{7.24}$$

Note that (7.24) becomes the usual discretization in the case $a(x) = 1$.

For slowly varying $a(x)$, we can use a standard multigrid scheme much like the algorithm used for the model problem. One reasonable approach in this case is to use linear interpolation for I_{2h}^h, full weighting for I_h^{2h}, and the following stencil for A^{2h}:

$$A^{2h} = \frac{1}{(2h)^2}\left(-a_{j-\frac{1}{2}}^{2h} \quad a_{j-\frac{1}{2}}^{2h} + a_{j+\frac{1}{2}}^{2h} \quad -a_{j+\frac{1}{2}}^{2h}\right), \quad 1 \le j \le \frac{n}{2}-1, \tag{7.25}$$

where the coefficients are determined by the simple averages

$$a^{2h}_{j+\frac{1}{2}} = \frac{a^h_{2j+\frac{1}{2}} + a^h_{2j+\frac{3}{2}}}{2}, \quad 0 \le j \le \frac{n}{2} - 1. \tag{7.26}$$

This same stencil can be obtained (Exercise 19) from the Galerkin principle

$$A^{2h} = I^{2h}_h A^h I^h_{2h}.$$

The performance of such a *standard* multigrid scheme will degrade as $a(x)$ begins to vary significantly from cell to cell. One way to understand how this degradation happens is to compare (7.24), the uniform mesh discretization of the variable-coefficient problem, to (7.18)–(7.19), the variable-mesh discretization of the constant coefficient Poisson problem. From the perspective of the fine-grid point x^h_{2j+1}, the variable coefficient problem can be viewed as a Poisson problem with variable mesh sizes defined by

$$h_{2j+\frac{1}{2}} = \frac{\gamma^h_{2j+1}}{a^h_{2j+\frac{1}{2}}}, \quad h_{2j+\frac{3}{2}} = \frac{\gamma^h_{2j+1}}{a^h_{2j+\frac{3}{2}}}, \tag{7.27}$$

where

$$\gamma^h_{2j+1} = h \sqrt{\frac{2a^h_{2j+\frac{1}{2}} a^h_{2j+\frac{3}{2}}}{a^h_{2j+\frac{1}{2}} + a^h_{2j+\frac{3}{2}}}}, \tag{7.28}$$

(Exercise 20). The point that we should get from these rather cumbersome formulas is that the variable-coefficient problem on a uniform mesh can be viewed locally as a constant-coefficient equation on a nonuniform mesh. This relationship is important because it says that standard multigrid for the variable-coefficient problem corresponds to using interpolation based on simple averaging for the nonuniform mesh problem. If the point x^h_{2j+1} of the variable mesh is close to x^h_{2j}, but far from x^h_{2j+2}, then simple averaging cannot accurately represent smooth components (Exercise 21).

A simple remedy is to define interpolation for the variable-coefficient problem guided by the variable-mesh case. The easiest way is to use operator interpolation, which gives (Exercise 22)

$$v^h_{2j} = v^{2h}_j, \tag{7.29}$$

$$v^h_{2j+1} = \frac{a^h_{2j+\frac{1}{2}} v^{2h}_j + a^h_{2j+\frac{3}{2}} v^{2h}_{j+1}}{a^h_{2j+\frac{1}{2}} + a^h_{2j+\frac{3}{2}}}, \quad 1 \le j \le \frac{n}{2} - 1.$$

Restriction and the coarse-grid operator can then be defined by the variational relations

$$I^{2h}_h = \frac{1}{2} \left(I^h_{2h} \right)^T \quad \text{and} \quad A^{2h} = I^{2h}_h A^h I^h_{2h}.$$

Variable-coefficient problems can be handled in two dimensions by an analogous, but more complicated, operator interpolation scheme. The principle difficulty in extending the one-dimensional approach is that the two-dimensional version of (7.21)

$a(x) = 1 + \rho \sin(k\pi x)$						$a(x) = 1 + \rho \, \mathrm{rand}(x)$
ρ	$k = 3$	$k = 25$	$k = 50$	$k = 100$	$k = 200$	$k = 400$	
0	0.085	0.085	0.085	0.085	0.085	0.085	0.085
0.25	.084	.098	.098	.094	.093	.083	0.083
0.50	.093	.185	.194	.196	.195	.187	0.173
0.75	.119	.374	.387	.391	.390	.388	0.394
0.85	.142	.497	.510	.514	.514	.526	0.472
0.95	.191	.681	.690	.694	.699	.745	0.672

Table 7.4: *Average convergence factors for 20 V-cycles of the variable-coefficient method. With $a(x) = 1 + \rho \sin(k\pi x)$, there is strong dependence on ρ, but relative insensitivity to all but the smallest value of k. With $a(x) = 1 + \rho \, \mathrm{rand}(x)$, convergence factors depend strongly on the amplitude ρ.*

generally cannot be solved for the error corresponding to the fine-grid point in terms of neighboring coarse-grid points alone. For example, for the nine-point stencil

$$A_{i,j}^h = \frac{1}{3h^2} \begin{pmatrix} -1 & -1 & -1 \\ -1 & 8 & -1 \\ -1 & -1 & -1 \end{pmatrix} \tag{7.30}$$

with a fully coarsened grid (i.e., every other fine-grid line is deleted), the equation at any fine-grid point involves at least four points that do not correspond to coarse-grid points. This difficulty can be reduced by considering semicoarsening algorithms, but the equations at fine-grid points still involve other fine-grid points that do not belong to the coarse grid. Nevertheless, there are effective ways to deal with this difficulty that involve collapsing the stencil to eliminate these troublesome fine-grid couplings, which in turn allow approximate operator interpolation. We leave this issue to the study of algebraic multigrid in the next chapter.

Numerical example. We illustrate the performance of a basic multigrid method applied to two variable-coefficient problems. We solve the one-dimensional problem (7.23) using the discretization in (7.24) and coarse-grid operator (7.25). The variable coefficient is

$$a(x) = 1 + \rho \sin(k\pi x), \tag{7.31}$$

for various choices of $\rho > 0$ and positive integers k, and

$$a(x) = 1 + \rho \, \mathrm{rand}(x), \tag{7.32}$$

where $\mathrm{rand}(x)$ returns a random number between –1 and 1. Noting that $a(x) = 1$ corresponds to the Poisson equation, these functions correspond to perturbations of the model problem. By increasing the size of the perturbation and the amount of local change in the perturbation, we are able to explore the behavior of the method for a wide variety of situations.

The grid size in all tests is $n = 1024$, which we found to be representative of the tests done over a wide range of n . We use V(2,1)-cycles based on Gauss–Seidel relaxation, full weighting, and linear interpolation. The results are displayed in Table 7.4.

For coefficients (7.31), the method depends strongly on the value of ρ, but very little on the value of the wavenumber k. For $\rho = 0.25$, the method works very well, with results akin to the model Poisson problem. However, with increasing ρ, performance degrades rapidly until convergence becomes quite poor. This dependence on ρ is to be expected: for ρ near 1, $a(x)$ has significant variation, from values near 0 to values near 1. One might expect that performance would also depend strongly on the wavenumber k. However, apart from the noticeable jump in convergence factors between $k = 3$ and $k = 25$, this does not appear to be the case, suggesting that standard multigrid methods are more sensitive to the amplitude of coefficient variations than to their frequency.

For coefficient (7.32), we drop the regular oscillatory variation in $a(x)$ in favor of random jumps. The various amplitudes, ρ, of the coefficient match those in the first case. Here, we again see a strong correlation between the amplitude and the convergence factor. For small amplitudes, the method performs well, but the performance degrades as ρ nears 1. It is interesting that the method seems relatively unaffected by the random variations in the coefficient. It performs at least as well with the random coefficients as it does with all but the lowest wavenumber case ($k = 3$) of the smoothly varying coefficients. ◇◇

These examples serve to highlight the point that a basic multigrid scheme may be used quite effectively for the variable-coefficient problem, provided the variation in the coefficient function is not too drastic. On the other hand, widely varying coefficients require more sophisticated multigrid methods. One approach to such problems is to use an algebraic multigrid method, which is the subject of the next chapter.

Exercises

Neumann Boundary Conditions

1. **Differential nonuniqueness.** Show that Neumann problem (7.1) does not have a *unique* solution by showing that if u solves (7.1), then $u + c$ is also a solution for any constant c. This means that if *any* particular solution exists, then another can be constructed simply by adding a constant. Now show that adding a general constant produces *all* solutions of (7.1). Hint: If u and v solve (7.1), then $(u - v)'' = 0$, so $u - v = ax + c$ for some constants a and c; now apply the boundary conditions.

2. **Differential solvability.** Show that Neumann problem (7.1) is solvable only for the special source terms f that satisfy integral compatibility condition (7.4). Hint: Assume u solves (7.1), then integrate both sides of the equation and apply the boundary conditions.

3. **Discrete null space.** Show by inspecting (7.3) that the null space of A^h is the set $\{c1^h\}$.

4. **Discrete nonuniqueness.** Show that discrete system (7.2)–(7.3) does not have a *unique* solution by proving that if $\mathbf{u}^h$ is a particular solution, then the general solution is $\mathbf{u}^h + c1^h$. Hint: First use Exercise 3 to confirm that $\mathbf{u}^h + c1^h$ is indeed a solution; then prove that any two solutions must differ by a vector in the null space of A^h.

5. **Discrete solvability.** Show that discrete system (7.9)–(7.10) is solvable only for the special source terms $\hat{\mathbf{f}}^h$ that satisfy discrete compatibility condition (7.7). Hint: Use Exercise 3, the fact that the range of a matrix is the orthogonal complement of the null space of its transpose, and the symmetry of $\hat{A}^h$.

6. **Restriction at the boundary points.** Using the interpolation operator at the left boundary given in (7.12) and the variational condition $I_h^{2h} = \frac{1}{2}\left(I_{2h}^h\right)^T$, derive the restriction formula at the left boundary point, $f_0^{2h} \leftarrow \frac{1}{2}f_0^h + \frac{1}{4}f_1^h$.

7. **Coarse-grid solvability.** Consider the coarse-grid equation

$$A^{2h}\mathbf{u}^{2h} = I_h^{2h}\left(\mathbf{f}^h - A^h\mathbf{u}^h\right).$$

(a) Show that interpolation preserves constants; that is, $\mathbf{1}^h = I_{2h}^h\mathbf{1}^{2h}$.

(b) Show that the variational property, $I_{2h}^h = c(I_h^{2h})^T$, guarantees that the discrete compatibility condition (7.7) is satisfied (that is, the coarse-grid problem is solvable).

(c) The variational property is actually not necessary for the discrete compatibility condition to hold. Show that it is enough for restriction to satisfy the property that its column sums equal a given constant γ (this means that a fine-grid residual is distributed to coarse-grid points with weights that sum to γ).

8. **Inhomogeneous Neumann conditions.** Consider the two-point boundary value problem

$$
\begin{aligned}
-u''(x) &= f(x), \quad 0 < x < 1, \\
u'(0) &= g_0, \\
u'(1) &= g_1,
\end{aligned}
$$

where g_0 and g_1 are given constants. Modify the essential concepts developed for (7.1) to accommodate this inhomogeneous case. In particular, show that

(a) compatibility condition (7.4) now involves a nonzero right side;

(b) the right side f^h of (7.2) and $\hat{f}_h$ of (7.9) change accordingly to incorporate g_0 and g_1;

(c) the rest of the development (the uniqueness condition of (7.5) and (7.10), the discrete compatibility condition in (7.7), and the coarse-grid correction process) is essentially unchanged.

Anisotropic Problems

9. **Two sources of anisotropy.** Show that, after discretization of the two-dimensional Poisson equation (7.14), the small parameter $\epsilon > 0$ in front of the u_{yy} term is equivalent to using a mesh spacing of $h_x = h$ in the x-direction and mesh spacing $h_y = \frac{h}{\sqrt{\epsilon}}$ in the y-direction.

10. **Eigenvalues for point Jacobi method.** Suppose the weighted point Jacobi method is applied to the system of equations corresponding to (7.15). Recall that A^h can be expressed as $D - L - U$, where D represents the diagonal elements of A^h and U and L represent the respective upper and lower triangular parts of A^h. Then the weighted Jacobi iteration matrix is given by $P_J = I - \omega D^{-1} A^h$.

(a) Write out a typical equation of the eigenvalue system $P_J \mathbf{v} = \lambda \mathbf{v}$.

(b) Assume an $n \times n$ grid and an eigenvector solution of the form

$$v_{ij} = \sin\left(\frac{ik\pi}{n}\right) \sin\left(\frac{j\ell\pi}{n}\right), \quad 1 \le k, \ell \le n - 1.$$

Using sine addition rules, simplify this eigenvalue equation, cancel common terms, and show that the eigenvalues are given by

$$\lambda_{k\ell} = 1 - \frac{2\omega}{1 + \epsilon}\left(\sin^2\left(\frac{k\pi}{2n}\right) + \epsilon \sin^2\left(\frac{\ell\pi}{2n}\right)\right), \quad 1 \le k, \ell \le n.$$

11. **Eigenvalues for block Jacobi method.** Consider A^h given in (7.16) and write it in the form $A^h = \mathcal{D} - U - L$, where $\mathcal{D}$ is the block diagonal matrix consisting of the blocks D on the diagonal and U and L are the respective lower and upper triangular parts of A^h.

(a) Show that the weighted block Jacobi iteration matrix can be written in the form

$$P_J = \mathcal{D}^{-1}(U + L) = I - \omega \mathcal{D}^{-1} A^h.$$

(b) Write out a typical equation of the eigenvalue system $P_J \mathbf{v} = \lambda \mathbf{v}$.

(c) Assume an $n \times n$ grid and an eigenvector solution of the form

$$v_{ij} = \sin\left(\frac{ik\pi}{n}\right) \sin\left(\frac{j\ell\pi}{n}\right), \quad 1 \le k, \ell \le n - 1.$$

Using sine addition rules, simplify this eigenvalue equation, cancel common terms, and show that the eigenvalues are given by

$$\lambda_{k,\ell} = 1 - \frac{2\omega}{2\sin^2\left(\frac{k\pi}{2n}\right) + \epsilon}\left(\sin^2\left(\frac{k\pi}{2n}\right) + \epsilon \sin^2\left(\frac{\ell\pi}{2n}\right)\right), \quad 1 \le k, \ell \le n.$$

(d) Compare the magnitudes of the eigenvalues of the block Jacobi method to those of the point Jacobi method (previous problem).

12. **Semicoarsening/point relaxation complexity.** Using the techniques of Chapter 4, for an $n \times n$ grid, show that with semicoarsening and point relaxation, the computational cost of a W-cycle is larger than $O(n^2)$, but that a V-cycle retains $O(n^2)$ complexity.

13. **Semicoarsening/line relaxation complexity.** Modifying the argument of the previous problem, show that with semicoarsening and line relaxation, the computational cost of a V-cycle is $O(n^2)$.

Variable-Mesh and Variable-Coefficient Problems

14. **Variable-mesh discretization.** Show that when model problem (7.17) is discretized using second-order finite difference approximations, difference equations (7.18) and (7.19) result.

15. **Linear interpolation for variable meshes.** Show that linear interpolation on a nonuniform grid leads to the formula expressed in (7.20). Hint: Construct a linear function in the interval $[x_j^{2h}, x_{j+1}^{2h}]$ that equals $\mathbf{v}^{2h}$ at the end points, then evaluate it at $x_{2j+1}^h = x_j^{2h} + h_{2j+\frac{1}{2}}$.

16. **Operator interpolation for variable meshes.** Show that interpolation formulas (7.20) and (7.22) are algebraically equivalent.

17. **Galerkin operator for variable meshes.** Show that the Galerkin principle applied to problem (7.18)–(7.19) leads to a similar, but generally different stencil. Show that the Galerkin principle does hold if (7.18) is multiplied on both sides by $(h_{j-\frac{1}{2}} + h_{j+\frac{1}{2}})/2$.

18. **Variable-coefficient discretization.** Verify that the discretization of the variable-coefficient problem (7.23) results in the difference equation (7.24).

19. **Galerkin operator for variable coefficients.** Show that the Galerkin coarse-grid operator for (7.24), using linear interpolation and full weighting, is the same as operator (7.25)–(7.26) obtained by simple averages of the fine-grid coefficients.

20. **Variable-coefficient versus variable-mesh.** Verify formula (7.27)–(7.28) that relates the variable-coefficient problem on a uniform mesh to Poisson's equation on a nonuniform mesh.

21. **Simple averaging for variable meshes.** To see that interpolation defined by simple averages can be ineffective for grids with widely varying mesh sizes, consider the three grid points $x_0^h = x_0^{2h} = 0, x_1^h = \epsilon h$, and $x_2^h = x_1^{2h} = 2h$, where h and ϵ are small positive parameters. Note that the smooth linear function $u(x) = \frac{x}{2h}$ is 0 at x_0^h and 1 at x_2^h. Show that it cannot be approximated well at x_1^h by the average of these end point values. Obtain an expression for the error at x_1^h in terms of h and ϵ and discuss the behavior as $\epsilon \to 0$.

22. **Interpolation for variable-coefficient problems.** Use the operator interpolation approach (assume the error satisfies the residual equation) to derive interpolation formulas (7.29).

Chapter 8

Algebraic Multigrid (AMG)

A natural question arises: Can we apply multigrid techniques when there is no *grid*? Suppose we have relationships among the unknowns that are similar to those in the model problem, but the physical locations of the unknowns are themselves unknown (or immaterial). Can we hope to apply the tools we have developed? A related question is: Can we apply multigrid in the case where grid locations are known but may be highly unstructured or irregular, making the selection of a coarse grid problematic? These are the problems that are addressed by a technique known as *algebraic multigrid*, or AMG [6]. We describe AMG using many of the concepts and principles developed in [18]. For a theoretical foundation of general algebraic methods, of which AMG is a part, see [5].

For any multigrid algorithm, the same fundamental components are required. There must be a sequence of grids, intergrid transfer operators, a relaxation (smoothing) operator, coarse-grid versions of the fine-grid operator, and a solver for the coarsest grid.

Let us begin by deciding what we mean by a *grid*. Throughout this chapter, we look to standard multigrid (which we refer to as the *geometric case*) to guide us in defining AMG components. In the geometric case, the unknown variables u_i are defined at known spatial locations (grid points) on a fine grid. We then select a subset of these locations as a coarse grid. As a consequence, a subset of the variables u_i is used to represent the solution on the coarse grid. For AMG, by analogy, we seek a subset of the variables u_i to serve as the coarse-grid unknowns. A useful point of view, then, is to identify the grid points with the indices of the unknown quantities. Hence, if the problem to be solved is $A\mathbf{u} = \mathbf{f}$ and

$$\mathbf{u} = \begin{bmatrix} u_1 \\ u_2 \\ \vdots \\ u_n \end{bmatrix},$$

then the fine-grid points are just the indices $\{1, 2, \ldots, n\}$.

Having defined the grid points, the connections within the grid are determined by the *undirected adjacency graph* of the matrix A. Letting the entries of A be a_{ij}, we associate the vertices of the graph with the grid points and draw an edge between the ith and jth vertices if either $a_{ij} \neq 0$ or $a_{ji} \neq 0$. The connections in

$$A = \begin{bmatrix} X & X & & X & X & \\ X & X & X & X & & \\ & X & X & X & X & X \\ X & X & X & X & & \\ X & & X & & X & X \\ & X & & & X & X \end{bmatrix}$$

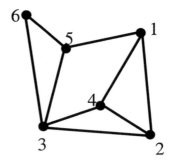

Figure 8.1: *The nonzero structure of A, where X indicates a nonzero entry, is shown on the left. The resulting undirected adjacency graph appears on the right.*

the grid are the edges in the graph; hence, the grid is entirely defined by the matrix A. A simple example of this relationship is given in Fig. 8.1.

Now that we can represent the fine grid, how do we select a coarse grid? With standard multigrid methods, smooth functions are *geometrically* or physically smooth; they have a low spatial frequency. In these cases, we assume that relaxation smooths the error and we select a coarse grid that represents smooth functions accurately. We then choose intergrid operators that accurately transfer smooth functions between grids.

With AMG, the approach is different. We first select a relaxation scheme that allows us to determine the nature of the smooth error. Because we do not have access to a physical grid, the sense of smoothness must be defined algebraically. The next step is to use this sense of smoothness to select coarse grids, which will be subsets of the unknowns. A related issue is the choice of intergrid transfer operators that allow for effective coarsening. Finally, we select the coarse-grid versions of the operator A, so that coarse-grid correction has the same effect that it has in geometric multigrid: it must eliminate the error components in the range of the interpolation operator.

Algebraic Smoothness

Having chosen a relaxation scheme, the crux of the problem is to determine what is meant by smooth error. If the problem provides no geometric information (for example, true grid point locations are unknown), then we cannot simply examine the Fourier modes of the error. Instead, we must proceed by analogy. In the geometric case, the most important property of smooth error is that it is not effectively reduced by relaxation. Thus, we now *define* smooth error loosely to be any error that is not reduced effectively by relaxation.

That was simple. Of course, we still need to figure out exactly what this definition means. To do this in the simplest case, we focus on weighted point Jacobi relaxation. We also assume that A is a symmetric *M-matrix*: it is symmetric ($A^T = A$) and positive-definite ($\mathbf{u}^T A \mathbf{u} > 0$ for all $\mathbf{u} \neq \mathbf{0}$) and has positive diagonal entries and nonpositive off-diagonal entries. These properties are shared by matrices arising from the discretization of many (not all) scalar elliptic differential equations. These assumptions are not necessary for AMG to work. However, the original theory of AMG was developed for symmetric M-matrices, and if A is far from being an M-matrix, it is less likely that standard AMG will be effective in solving the problem.

Recall from Chapter 2 that the weighted point Jacobi method can be expressed as

$$\mathbf{v} \leftarrow \mathbf{v} + \omega D^{-1}(f - A\mathbf{v}),$$

where D is the diagonal of A. As in previous chapters, $\mathbf{v}$ is a computed approximation to the exact solution $\mathbf{u}$. Remember that the error propagation for this iteration can be written as

$$\mathbf{e} \leftarrow \left(I - \omega D^{-1}A\right)\mathbf{e}. \tag{8.1}$$

Weighted Jacobi relaxation, as we know, has the property that after making great progress toward convergence, it stalls, and little improvement is made with successive iterations. At this point, we define the error to be *algebraically smooth*. It is useful to examine the implications of algebraic smoothness. Because all of AMG is based on this concept, the effort is worthwhile.

By our definition, algebraic smoothness means that the size of $\mathbf{e}^{i+1}$ is not significantly less than that of $\mathbf{e}^i$. We need to be more specific about the concept of size. A natural choice is to measure the error in the A-norm, which is induced by the A-inner product. As defined in Chapter 5, we have

$$\|\mathbf{e}\|_A = (A\mathbf{e}, \mathbf{e})^{1/2}.$$

Using this norm and (8.1), we see that an algebraically smooth error is characterized by

$$\| \left(I - \omega D^{-1}A\right)\mathbf{e}\|_A \approx \|\mathbf{e}\|_A.$$

When we assume that $\omega = \alpha \|D^{-1/2}AD^{-1/2}\|^{-1}$ for some fixed $\alpha \in (0, 2)$ and that $\|D^{-1/2}AD^{-1/2}\|$ is $O(1)$ (for the model problem, it is bounded by 2), it can be shown (Exercise 1) that

$$(D^{-1}A\mathbf{e}, A\mathbf{e}) \ll (\mathbf{e}, A\mathbf{e}).$$

Writing this expression in components yields

$$\sum_{i=1}^{n} \frac{r_i^2}{a_{ii}} \ll \sum_{i=1}^{n} r_i e_i.$$

This implies that, *at least on average*, algebraically smooth error $\mathbf{e}$ satisfies

$$|r_i| \ll a_{ii}|e_i|.$$

We write this condition loosely as

$$A\mathbf{e} \approx \mathbf{0} \tag{8.2}$$

and read it as meaning that smooth error has relatively small residuals. We will appeal to this condition in the development of the AMG algorithm. While our analysis here is for weighted Jacobi, Gauss–Seidel relaxation is more commonly used for AMG. A similar, though slightly more complicated analysis can be performed for Gauss–Seidel relaxation and also leads to condition (8.2) (Exercise 11).

One immediate implication of (8.2) is that $r_i \approx 0$, so

$$a_{ii}e_i \approx -\sum_{j \neq i} a_{ij}e_j; \tag{8.3}$$

that is, if $\mathbf{e}$ is a smooth error, then e_i can be approximated well by a weighted average of its neighbors. This fact gives us an important foothold in determining an interpolation operator.

A short digression here should serve to clarify the difference between algebraic and geometric smoothness. We consider a simple example due to Stüben in his introduction to AMG [22]. Suppose the problem

$$-au_{xx} - cu_{yy} + bu_{xy} = f(x, y), \tag{8.4}$$

with homogeneous Dirichlet boundary conditions, is discretized on the unit square using a uniform grid and the finite-difference stencils

$$D_{xx}^h = \frac{1}{h^2} \begin{pmatrix} 1 & -2 & 1 \end{pmatrix}, \qquad D_{yy}^h = \frac{1}{h^2} \begin{pmatrix} 1 \\ -2 \\ 1 \end{pmatrix},$$

$$D_{xy}^h = \frac{1}{2h^2} \begin{pmatrix} -1 & 1 \\ 1 & -2 & 1 \\ & 1 & -1 \end{pmatrix}. \tag{8.5}$$

Also suppose that the coefficients a, b, and c are locally constant, but have different values in different quadrants of the domain. Specifically, let the coefficients be defined in the square domain as shown below:

$a = 1$ $c = 1000$ $b = 0$	$a = 1$ $c = 1$ $b = 2$
$a = 1$ $c = 1$ $b = 0$	$a = 1000$ $c = 1$ $b = 0$

Note that this discretization does not produce an M-matrix.

Using a zero right side and a random initial guess, the norm of the error essentially stops changing after eight sweeps of Gauss–Seidel. By our definition, this error is algebraically smooth. However, it does not appear to be smooth in the geometric sense (Fig. 8.2). In fact, in three of the four quadrants it is geometrically quite oscillatory! But because the iteration has stalled, this is precisely the error AMG must account for in coarse-grid correction. We return to this example later.

Influence and Dependence

Most of AMG rests on two fundamental concepts. We have just discussed the first concept, namely, smooth error. The second important concept is that of *strong dependence* or *strong influence*. Because of the dominance of the diagonal entry (A is an M-matrix), we associate the ith equation with the ith unknown; the job of the ith equation is to determine the value of u_i. Of course, it usually takes all of

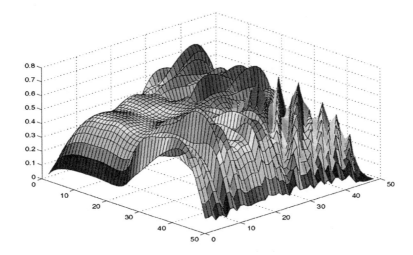

Figure 8.2: *Error that is algebraically smooth, but not geometrically smooth for* (8.4).

the equations to determine any given variable precisely. Nevertheless, our first task is to determine which *other* variables are most important in the ith equation; that is, which u_j are most important in the ith equation in determining u_i?

One answer to this question lies in the following observation: if the coefficient, a_{ij}, which multiplies u_j in the ith equation, is large relative to the other coefficients in the ith equation, then a small change in the value of u_j has more effect on the value of u_i than a small change in other variables in the ith equation. Intuitively, it seems logical that a variable whose value is instrumental in determining the value for u_i would be a good value to use in the interpolation of u_i. Hence, such a variable (point) should be a candidate for a coarse-grid point. This observation suggests the following definition.

Definition 1. Given a threshold value $0 < \theta \le 1$, the variable (point) u_i *strongly depends* on the variable (point) u_j if

$$-a_{ij} \ge \theta \max_{k \ne i} \{-a_{ik}\}. \qquad (8.6)$$

This says that grid point i strongly depends on grid point j if the coefficient a_{ij} is comparable in magnitude to the largest off-diagonal coefficient in the ith equation. We can state this definition from another perspective.

Definition 2. If the variable u_i strongly depends on the variable u_j, then the variable u_j *strongly influences* the variable u_i.

With the twin concepts of smooth error and strong influence/dependence in hand, we can return to the task of defining the multigrid components for AMG. As with any multigrid algorithm, we begin by defining a two-grid algorithm, then proceed to multigrid by recursion. Having defined the relaxation scheme, we have several tasks before us:

- select a coarse grid so that the smooth components can be represented accurately;

- define an interpolation operator so that the smooth components can be accurately transferred from the coarse grid to the fine grid; and

- define a restriction operator and a coarse-grid version of A using the variational properties.

Defining the Interpolation Operator

Assume for the moment that we have already designated the coarse-grid points. This means that we have a partitioning of the indices $\{1, 2, \ldots, n\} = C \cup F$, where the variables (points) corresponding to $i \in C$ are the coarse-grid variables. These coarse-grid variables are also fine-grid variables; the indices $i \in F$ represent those variables that are *only* fine-grid variables. Next, suppose that $e_i, i \in C$, is a set of values on the coarse grid representing a smooth error that must be interpolated to the fine grid, $C \cup F$. What do we know about e_i that allows us to build an interpolation operator that is accurate? With geometric multigrid, we use linear interpolation between the coarse grid points. With an unstructured, or perhaps nonexistent, grid, the answer is not so obvious.

If a C-point j strongly influences an F-point i, then the value e_j contributes heavily to the value of e_i in the ith (fine-grid) equation. It seems reasonable that the value e_j in the coarse-grid equation could therefore be used in an interpolation formula to approximate the fine-grid value e_i. This idea can be strengthened by noting that the following bound must hold for smooth error on average, that is, for most i (Exercise 2):

$$\sum_{j \neq i} \left(\frac{|a_{ij}|}{a_{ii}} \right) \left(\frac{e_i - e_j}{e_i} \right)^2 \ll 1, \quad 1 \leq i \leq n. \tag{8.7}$$

The left side of the inequality is a sum of products of nonnegative terms. These products must be very small, which means that one or both of the factors in each product must be small. But if e_i strongly depends on e_j, we know that $-a_{ij}$ could be comparable to a_{ii}. Therefore, for these strongly influencing e_j's, it must be true that $e_i - e_j$ is small; that is, $e_j \approx e_i$. We describe this by saying that *smooth error varies slowly in the direction of strong connection*. Thus, we have a justification for the idea that the fine-grid quantity u_i can be interpolated from the coarse-grid quantity u_j if i strongly depends on j.

For each fine-grid point i, we define N_i, the *neighborhood of i*, to be the set of all points $j \neq i$ such that $a_{ij} \neq 0$. These points can be divided into three categories:

- the neighboring coarse-grid points that strongly influence i; this is the *coarse interpolatory set* for i, denoted by C_i;

- the neighboring fine-grid points that strongly influence i, denoted by D_i^s; and

- the points that do not strongly influence i, denoted by D_i^w; this set may contain both coarse- and fine-grid points; it is called the set of *weakly connected neighbors*.

The goal is to define the *interpolation* operator I_{2h}^h (although physical grids may not be present, we continue to denote fine-grid quantities by h and coarse-grid quantities by $2h$). We require that the ith component of $I_{2h}^h \mathbf{e}$ be given by

$$(I_{2h}^h \mathbf{e})_i = \begin{cases} e_i & \text{if } i \in C, \\ \displaystyle\sum_{j \in C_i} \omega_{ij} e_j & \text{if } i \in F, \end{cases} \tag{8.8}$$

where the interpolation weights, ω_{ij}, must now be determined.

Recall that the main characteristic of smooth error is that the residual is small: $\mathbf{r} \approx \mathbf{0}$. We can write the ith component of this condition as

$$a_{ii} e_i \approx -\sum_{j \in N_i} a_{ij} e_j.$$

Splitting the sum into its component sums over the coarse interpolatory set, C_i, the fine-grid points with strong influence, D_i^s, and the weakly connected neighbors, D_i^w, we have

$$a_{ii} e_i \approx -\sum_{j \in C_i} a_{ij} e_j - \sum_{j \in D_i^s} a_{ij} e_j - \sum_{j \in D_i^w} a_{ij} e_j. \tag{8.9}$$

To determine the ω_{ij}, we need to replace the e_j in the second and third sums on the right side of (8.9) with approximations in terms of e_i or e_j, where $j \in C_i$.

Consider the third sum over points that are weakly connected to point i. We distribute these terms to the diagonal coefficient; that is, we simply replace e_j in the rightmost sum by e_i, giving

$$\left(a_{ii} + \sum_{j \in D_i^w} a_{ij} \right) e_i \approx -\sum_{j \in C_i} a_{ij} e_j - \sum_{j \in D_i^s} a_{ij} e_j. \tag{8.10}$$

We can justify this distribution in the following way. Suppose we have underestimated the dependence, so that e_i does depend strongly on the value of the points in D_i^w. Then the fact that smooth error varies *slowly* in the direction of strong dependence means that $e_i \approx e_j$ and the distribution to the diagonal makes sense. Alternatively, suppose the value of e_i does not depend strongly on the points in D_i^w. Then the corresponding value if a_{ij} will be small and any error committed in making this assignment will be relatively insignificant.

Treating the second sum over D_i^s is a bit more complicated because we must be more careful with these strong connections. We might simply distribute these terms to the diagonal, and, indeed, this would work nicely for many problems. However, experience has shown that it is better to distribute the terms in D_i^s to C_i. Essentially, we want to approximate the e_j's in this sum with weighted sums of the e_k for $k \in C_i$. That is, we want to replace each e_j, where j is a fine-grid point that strongly influences i, with a linear combination of values of e_k from the coarse interpolatory set of the point i. We do this, for each fixed $j \in D_i^s$, by making the approximation

$$e_j \approx \frac{\displaystyle\sum_{k \in C_i} a_{jk} e_k}{\displaystyle\sum_{k \in C_i} a_{jk}}. \tag{8.11}$$

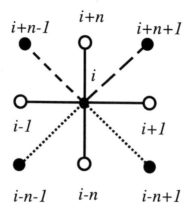

Figure 8.3: *Strong and weak influences on the point i. Coarse-grid points are shown as open circles. C-points with a strong influence on i are indicated with solid lines, F-points that strongly influence i with dashed lines, and F-points that weakly influence i with dotted lines.*

The numerator is appropriate because the e_j are strongly influenced by the e_k in proportion to the matrix entries a_{jk}. The denominator is chosen to ensure that the approximation interpolates constants exactly. Notice that this approximation *requires* that if i and j are any two strongly connected fine-grid points, then they must have at least one point common to their coarse interpolatory sets C_i and C_j.

If we now substitute (8.11) into (8.10) and engage in a spate of algebra (Exercise 3), we find that the interpolation weights are given by

$$\omega_{ij} = -\frac{a_{ij} + \displaystyle\sum_{m \in D_i^s}\left(\frac{a_{im}a_{mj}}{\displaystyle\sum_{k \in C_i} a_{mk}}\right)}{a_{ii} + \displaystyle\sum_{n \in D_i^w} a_{in}}. \tag{8.12}$$

The calculation of the interpolation weights can be illustrated with a simple example. Consider the operator A, defined on a uniform $n \times n$ grid, by the stencil

$$\begin{bmatrix} -\frac{1}{2} & -2 & -\frac{1}{2} \\ -1 & \frac{29}{4} & -1 \\ -\frac{1}{8} & -2 & -\frac{1}{8} \end{bmatrix}.$$

Assume that the partition of the grid into the C- and F-points corresponds to *red-black* coarsening. For a typical interior fine-grid point i, the four points directly north, south, east, and west are the coarse interpolatory set C_i. Using $\theta = 0.2$ as the dependence threshold, the strong and weak influences on i are shown in Fig. 8.3.

The points northwest and northeast of i form the set D_i^s (fine-grid points with strong influence), while the points to the southeast and southwest form the set D_i^w

(points with weak influence). For this example, (8.9) becomes

$$\frac{29}{4}e_i = \underbrace{2e_{i+n} + 2e_{i-n} + e_{i-1} + e_{i+1}}_{C_i}$$
$$+ \underbrace{\frac{1}{2}u_{i+n-1} + \frac{1}{2}e_{i+n+1}}_{D_i^s}$$
$$+ \underbrace{\frac{1}{8}e_{i-n-1} + \frac{1}{8}e_{i-n+1}}_{D_i^w}.$$

Substituting e_i for the D_i^w points, e_{i-n+1} and e_{i-n-1}, in the rightmost sum yields

$$\left(\frac{29}{4} - \frac{1}{8} - \frac{1}{8}\right)e_i = \underbrace{2e_{i+n} + 2e_{i-n} + e_{i-1} + e_{i+1}}_{C_i \text{ points}} + \underbrace{\frac{1}{2}u_{i+n-1} + \frac{1}{2}e_{i+n+1}}_{D_i^s}. \quad (8.13)$$

Point i depends strongly on $i+n-1$, which is itself strongly dependent on points $i-1$ and $i+n$ in C_i. Similarly, i depends strongly on $i+n+1$, which in turn depends strongly on $i+1$ and $i+n$ from the coarse interpolatory set C_i. Using (8.11), we obtain the approximations

$$e_{i+n-1} \approx \frac{-2e_{i-1} - e_{i+n}}{-(2+1)}, \qquad e_{i+n+1} \approx \frac{-2e_{i+1} - e_{i+n}}{-(2+1)},$$

which we can substitute into the sum over the strong F-points in (8.13). After all the algebra is done, (8.13) reduces to the interpolation formula

$$e_i = \frac{7}{21}e_{i+n} + \frac{6}{21}e_{i-n} + \frac{4}{21}e_{i+1} + \frac{4}{21}e_{i-1}.$$

It is worth noting that, as in many good interpolation formulas, the coefficients are nonnegative and sum to one. A worthwhile exercise is to show that under appropriate conditions, this must be the case (Exercise 4).

Selecting the Coarse Grid

The preceding discussion of the interpolation operator assumed that we had already designated points of the coarse grid. We must now turn our attention to this critical task. We use the twin concepts of strong influence/dependence and smooth error, just as we did in defining interpolation. As in the geometric problem, we rely on the fundamental premise that the coarse grid must be one

- on which smooth error can be approximated accurately,

- from which smooth functions can be interpolated accurately, and

- that has substantially fewer points than the fine grid, so that the residual problem may be solved with relatively little expense.

The basic idea is straightforward. By examining the suitability of each grid point to be a point of one of the C_i sets, we make an initial partitioning of the grid points into C- and F-points. Then, as the interpolation operator is constructed, we make adjustments to this partitioning, changing points initially chosen as F-points to be C-points in order to ensure that the partitioning conforms to certain heuristic rules.

Before we can describe the coarsening process in detail, we need to make two more definitions and to introduce these heuristics. Denote by S_i the set of points that strongly influence i; that is, the points on which the point i strongly depends. Also denote by S_i^T the set of points that strongly depend *on* the point i. Armed with these definitions, we describe two heuristic criteria that guide the initial selection of the C-points:

H-1: For each F-point i, every point $j \in S_i$ that strongly influences i either should be in the coarse interpolatory set C_i or should strongly depend on at least one point in C_i.

H-2: The set of coarse points C should be a maximal subset of all points with the property that no C-point strongly depends on another C-point.

To motivate heuristic **H-1**, we examine the approximation (8.11) that was made in developing the interpolation formula. Recall that this approximation applies to points $j \in D_i^s$ that consist of F-points strongly influencing the F-point i. Because e_i depends on these points, their values must be represented in the interpolation formula in order to achieve accurate interpolation. But because they have not been chosen as C-points, they are represented in the interpolation formula only by distributing their values to points in C_i using (8.11). It seems evident that (8.11) will be more accurate if j is strongly dependent on several points in C_i. However, for the approximation to be made at all, j must be strongly dependent on at least one point in C_i. Heuristic **H-1** simply ensures that this occurs.

Heuristic **H-2** is designed to strike a balance on the size of the coarse grid. Multigrid efficiency is generally controlled by two properties: convergence factor and number of WUs per cycle. If the coarse grid is a large fraction of the total points, then the interpolation of smooth errors is likely to be very accurate, which, in turn, generally produces better convergence factors. However, relatively large coarse grids generally mean a prohibitively large amount of work in doing V-cycles. By requiring that no C-point strongly depends on another, **H-2** controls the size of the coarse grid because C-points tend to be farther apart. By requiring C to be a maximal subset (that is, no other point can be added to C without violating the ban on mutual strong dependence), **H-2** ensures that C is big enough to produce good convergence factors.

It is not always possible to enforce both **H-1** and **H-2** (see Exercise 5). Because the interpolation formula depends on **H-1** being satisfied, we choose to enforce **H-1** rigorously, while using **H-2** as a guide. While this choice may lead to larger coarse grids than necessary, experience shows that this trade-off between accuracy and expense is generally worthwhile.

The basic coarse-point selection algorithm proceeds in two passes. We first make an initial coloring of the grid points by choosing a preliminary partition into C- and F-points. The goal in the first pass is to create a set of C-points that have good approximation properties and also tend to satisfy **H-2**. Once the initial assignments have been made, we make a second pass, changing initial F-points to C-points as necessary to enforce **H-1**.

The Coloring Scheme

The first pass begins by assigning to each point i a measure of its potential quality as a C-point. There are several ways we can make this assessment, but the simplest is to count the number of other points strongly influenced by i. Because those points are the members of S_i^T, this count, λ_i, is the cardinality of S_i^T. Once the measures λ_i have been determined, we select a point with maximum λ_i value as the first point in C.

The point we just selected strongly influences several of the other points and should appear in the interpolation formula for each of them. This implies that the points that depend strongly on i should become F-points. We therefore assign all points in S_i^T to F, which is permissible because we already have a C-point, i, that strongly influences them. It is logical to look at *other* points that strongly influence these new F-points as potential C-points, because their values could be useful for accurate interpolations. Therefore, for each new F-point j in S_i^T, we increment the measure, λ_k, of each unassigned point k that strongly influences j; this would be each unassigned member of $k \in S_j$.

The process is then repeated. A new unassigned point i is found with maximum λ_i and it is assigned to C. The unassigned points $j \in S_i^T$ are then assigned to F and the measures of the unassigned points in S_j are incremented by 1. This process continues until all points have been assigned to C or F.

It is useful to observe that the coarsening determined by this method depends on several factors. Among the most influential is the order in which the grid points are scanned when seeking the next point with maximal λ. Because many, if not most, of the grid points will have the maximal value at the start, any of them could be selected as the first coarse point. Once the first point is selected, the rest proceeds as outlined. Again, any time there is more than one point having the maximal value, there are many possible coarsenings. The heuristics ensure that whatever specific coarse grid is obtained, it will have the desired properties: it provides a good representation of smooth error components, while keeping the size of the coarse grid reasonably small.

This coloring algorithm is best illustrated by an example. The upper left drawing in Fig. 8.4 shows the graph of the matrix for a nine-point stencil representing the Laplacian operator on a uniform grid. The operator stencil is

$$\frac{1}{h^2} \begin{pmatrix} -1 & -1 & -1 \\ -1 & 8 & -1 \\ -1 & -1 & -1 \end{pmatrix}. \tag{8.14}$$

For this example, the dependence threshold is immaterial; for any θ, every connection is one of strong dependence. Hence, each point strongly influences, and depends strongly upon, each of its neighbors. Initially, all interior points have a measure of $\lambda = 8$, all points along the side of the grid have a measure of $\lambda = 5$, and the four corner points of the grid have a measure of $\lambda = 3$. We also assume the points are stored in lexicographic order.

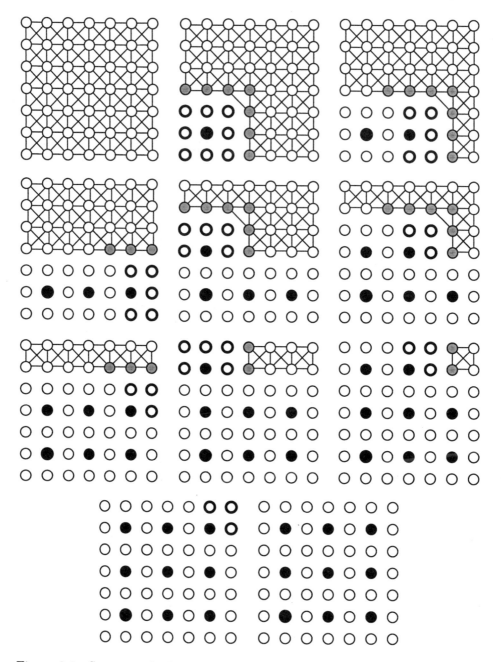

Figure 8.4: *Sequence of coloring steps for the nine-point Laplacian on a uniform grid. The upper left diagram is the original grid, the lower right the final coloring.*

The remaining drawings in Fig. 8.4 show the coloring of the grid as it evolves. At each step, newly designated C-points are shown in black; newly designated F-points are shown in white with a heavy border; and "undecided" neighbors of the new F-points whose λ values are updated are shaded in gray. Edges of the graph are removed as the algorithm accounts for the dependencies of the new C- and F-points. The lower right drawing of the figure displays the completed coloring.

It is useful to observe that for this example, coarse-grid selection is complete after this pass. A careful examination of the lower right drawing in the figure reveals that, because of the high connectivity of the graph, both **H-1** and **H-2** are satisfied by the C/F coloring produced. In addition, the coarse grid produced is exactly the standard full coarsening that one would use for a geometric multigrid method for this problem.

It is also instructive to examine the application of this coarsening to the five-point Laplacian stencil on a uniform grid. Once again, the first pass of the process not only produces a coloring satisfying both heuristics, but one that is very common in geometric solutions to the problem as well (Exercise 6).

It is not difficult to concoct an example that does not work so well. One example is illustrated in Fig. 8.5. Again, the nine-point Laplacian stencil (8.14) is used on an $n \times n$ uniform grid, but now we add the periodic boundary conditions $u_{i,j} = u_{i \pm n, j \pm n}$ with $n = 7$. Note that the initial measure is $\lambda = 8$ for all grid points. The thin lines indicate the extent of the grid with the first periodic replication shown outside the lines. At each step, newly designated C-points are shown in black; newly designated F-points are shown in white with a heavy border; and "undecided" neighbors of the new F-points whose λ values are updated are shaded in gray. Edges of the graph are removed as the algorithm accounts for the dependencies of the new C- and F-points. The final first-pass coloring (lower right of figure) violates **H-1**, with a large number of F-F dependencies between points not sharing a C-point.

This is an example of a problem for which it is impossible to satisfy both **H-1** and **H-2**. Because the coloring algorithm satisfies **H-2** and we have determined that we must satisfy **H-1**, a second pass for the coarsening algorithm must be done. In this pass, we examine each of the F-points in turn and determine if there are F-F dependencies with points not depending strongly on a common C-point. If this is the case, then one of the two F-points is tentatively changed into a C-point, and the examination moves on to the next F-point. When further F-F dependencies are encountered, we first attempt to satisfy **H-1** using points on the list of tentative points. The idea is to satisfy **H-1** by converting a minimal number of F-points into C-points.

Figure 8.6 displays the coarsening produced for the periodic nine-point Laplacian after the second pass of the coloring algorithm. The extra C-points are shaded black with a rim of dark gray. It can be seen that while the coloring now satisfies **H-1**, it no longer satisfies **H-2**. The coarse grid has a few more points, but the interpolation formula can be built for each F-point. The details of the second-pass algorithm are left as an exercise (Exercise 7).

Two examples highlight some important features of the coarsening algorithm. First, consider the problem

$$-u_{xx} - u_{yy} = f(x, y), \tag{8.15}$$

with homogeneous Dirichlet boundary conditions. Suppose (8.15) is discretized by finite elements using regular quadrilaterals of dimension $h_x \times h_y$. If $h_y / h_x \to 0$, then the finite element stencil approaches

$$\begin{pmatrix} -1 & -4 & -1 \\ 2 & 8 & 2 \\ -1 & -4 & -1 \end{pmatrix}.$$

While this stencil does not correspond to an M-matrix, the AMG coarsening algo-

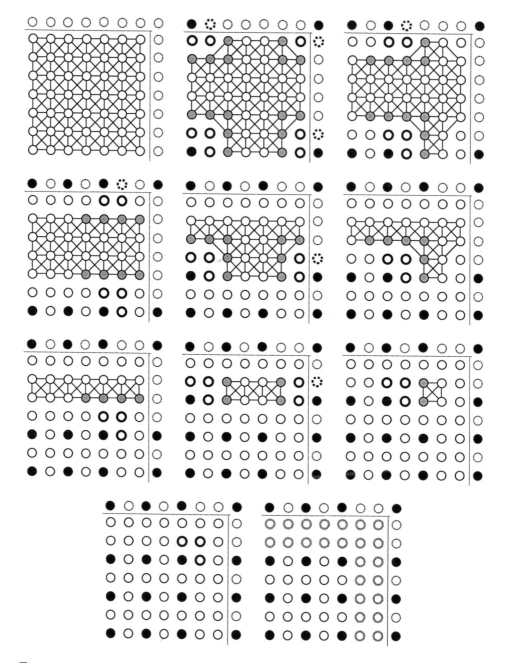

Figure 8.5: *Sequence of coloring steps for the nine-point Laplacian on a uniform grid with periodic boundary conditions. The upper left diagram is the original grid, the lower center the final coloring (first coloring pass). The gray circles in the lower right diagram are points that violate heuristic* **H-1**.

rithm performs quite well on this problem and is very instructive. Note that the problem is essentially equivalent to the discretization of

$$-\epsilon u_{xx} - u_{yy} = f,$$

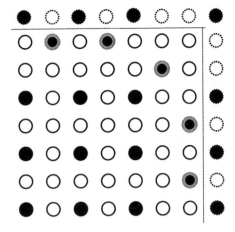

Figure 8.6: *Second coloring pass for the nine-point Laplacian problem with periodic boundary conditions. The added C-points are shown in the final coloring as black dots with thick gray outlines.*

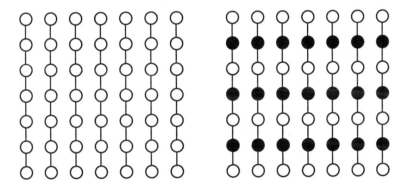

Figure 8.7: *Semi-coarsening produced for a stretched mesh (or anisotropic) problem. Shown on the left is the grid with only the strong dependencies indicated. On the right, the final coarsening is displayed.*

where ϵ is very small. The important feature is that the problem shows strong dependence in the y-direction, and little or no strong dependence in the x-direction. There are several subtle nuances associated with this discretization; our interest here, however, is in the coarse-grid selection.

The coloring produced for this problem is shown in Fig. 8.7. The algorithm generates a semicoarsened grid that is coarsened only in the y-direction. Again, this is precisely the coarse grid that would be used by a geometric semicoarsening method for this problem. Perhaps the most important observation is that the grid has been coarsened *only in the direction of strong dependence*. This makes sense: because smooth error varies slowly in the direction of strong dependence, interpolation can be performed accurately in that direction. AMG coarsening must exhibit this critical property.

Consider the nine-point Laplacian (Fig. 8.4) and the five-point Laplacian (Exercise 6). The coarsening for the five-point operator is called the red-black coarsening.

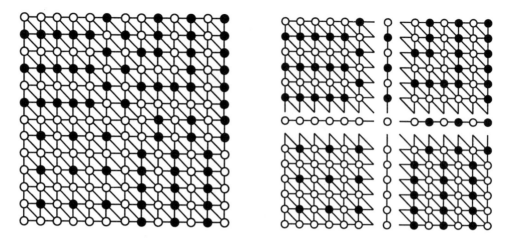

Figure 8.8: *Coarsening of problem* (8.4) *is shown on the left. Observe how the coarsening is in the direction of strong dependence in each of the four quadrants of the grid (right).*

The operator exhibits dependence in both the x- and y-directions, and the red-black coarsening is the only pattern that coarsens each grid line in both the x- and y-directions. The nine-point operator exhibits dependence in all directions: x, y, and diagonal directions. The resulting coarsening, known as *full coarsening*, coarsens in each of these directions. It is not possible to coarsen along x, y, and the diagonal grid line simultaneously; the algorithm instead coarsens every other line in each direction, with the intervening lines not represented on the coarse grid. The strong connectivity of this operator ensures that there are sufficient points available to interpolate the intervening lines well.

That AMG automatically coarsens in the directions of dependence is made apparent by the coarsening produced for the problem given in (8.4). The coarsening is shown in Fig. 8.8. Observe that in the upper left quadrant, where $a = 1$, $c = 1000$, and $b = 0$, the strong dependence is in the y-direction, and, in that region, the grid is coarsened primarily in the y-direction. Similarly, in the lower right quadrant where the coefficients are $a = 1000$, $c = 1$, and $b = 0$, the strong dependence is in the x-direction, and the grid is coarsened in the x-direction. In the lower left quadrant, the problem is just the normal Laplacian, and the coarsening is the standard full coarsening we observed for the nine-point Laplacian operator. Finally, in the upper right quadrant, where $a = 1$, $c = 1$, and $b = 2$, the coarsening is in the northwest/southeast direction, which is the direction of strong dependence for the stencil given in (8.5).

Coarse-Grid Operators

Recall that although physical grids may not be present, we continue to denote fine-grid quantities by h and coarse-grid quantities by $2h$. Once the coarse grid is chosen and the interpolation operator I_{2h}^h is constructed, the restriction operator

I_h^{2h} is defined using the usual variational property

$$I_h^{2h} = \left(I_{2h}^h\right)^T.$$

The coarse-grid operator is constructed using the Galerkin condition

$$A^{2h} = I_h^{2h} A^h I_{2h}^h. \tag{8.16}$$

The reason for defining interpolation and the coarse operator by these variational principle is that the resulting coarse-grid correction is optimal in the A^h-norm (Exercise 8).

Cycling Algorithms

We have now defined all the components necessary to create a two-grid correction algorithm for AMG: a relaxation scheme, a set of coarse-grid points C, a coarse-grid operator A^{2h}, and intergrid transfer operators I_h^{2h} and I_{2h}^h. Although we have discussed weighted Jacobi, Gauss–Seidel relaxation is often preferred. The two-grid correction algorithm appears exactly as it did for geometric multigrid, as shown below.

AMG Two-Grid Correction Cycle

$$\mathbf{v}^h \leftarrow \text{AMG}(\mathbf{v}^h, \mathbf{f}^h).$$

- Relax ν_1 times on $A^h \mathbf{u}^h = \mathbf{f}^h$ with initial guess $\mathbf{v}^h$.

- Compute the fine-grid residual $\mathbf{r}^h = \mathbf{f}^h - A^h \mathbf{v}^h$ and restrict it to the coarse grid by $\mathbf{r}^{2h} = I_h^{2h} \mathbf{r}^h$.

- Solve $A^{2h} \mathbf{e}^{2h} = \mathbf{r}^{2h}$ on Ω^{2h}.

- Interpolate the coarse-grid error to the fine grid by $\mathbf{e}^h = I_{2h}^h \mathbf{e}^{2h}$ and correct the fine-grid approximation by $\mathbf{v}^h \leftarrow \mathbf{v}^h + \mathbf{e}^h$.

- Relax ν_2 times on $A^h \mathbf{u}^h = \mathbf{f}^h$ with initial guess $\mathbf{v}^h$.

Having defined the two-grid correction algorithm, we can define other multigrid cycling schemes for AMG guided by geometric multigrid. For example, to create a V-cycle algorithm, we simply replace the direct solution of the coarse-grid problem with a recursive call to AMG on all grids except the coarsest grid, where we use a direct solver. W-cycles, μ-cycles, and FMG-cycles can also be created by strict analogy to the geometric multigrid case.

Costs of AMG

How expensive is the AMG algorithm? The geometric case has regular and predictable costs, both in terms of the storage and floating-point operation counts. By contrast, there are no *predictive* cost analyses available for AMG. This is because we do not know, in advance, the ratio of coarse- to fine-grid points. Furthermore, this ratio is unlikely to remain constant as the grid hierarchy is established. However, we have two simple tools that provide *a posteriori* cost estimates. Such estimates are useful in analyzing the performance of AMG and in building confidence that AMG is an effective tool for many problems. Additionally, they are extremely useful in debugging, code tuning, and algorithm development.

> **Definitions.** *Grid complexity* is the total number of grid points, on all grids, divided by the number of grid points on the finest grid. *Operator complexity* is the total number of nonzero entries, in all matrices A^{kh}, divided by the number of nonzero entries in the fine-grid operator A^h.

Thus, in the geometric case on the model problem, we have grid complexities of about 2, $\frac{4}{3}$, and $\frac{8}{7}$ for the one-, two-, and three-dimensional problems, respectively. Grid complexity gives an accurate measure of the storage required for the right sides and approximation vectors and can be directly compared to geometric multigrid.

Operator complexity indicates the total storage space required by the operators A^{kh} over all grids, which is generally not necessary in the geometric case. But it also has another use. Just as in the geometric case, the work in the solve phase of AMG is dominated by relaxation and residual computations, which are directly proportional to the number of nonzero entries in the operator. Hence, the work of a V-cycle turns out to be essentially proportional to the operator complexity. This proportionality is not perfect, but operator complexity is generally considered a good indication of the expense of the AMG V-cycle.

Performance

By its very nature, AMG is designed to apply to a broad range of problems, many far removed from the partial differential equations for which geometric multigrid was developed. The performance of AMG will naturally depend in large part on the peculiarities of the problem to which it is applied. Hence, we cannot hope to give a complete discussion of AMG performance. Accordingly, we close this chapter by discussing two numerical PDE examples.

Numerical example. Because it is helpful to illustrate the similarities and differences between the geometric and algebraic approaches, the first example is that of Chapter 4:

$$
\begin{aligned}
-u_{xx} - u_{yy} &= 2[(1 - 6x^2)y^2(1 - y^2) + (1 - 6y^2)x^2(1 - x^2)] && \text{in } \Omega, \\
u &= 0 && \text{on } \partial\Omega,
\end{aligned} \tag{8.17}
$$

where Ω is the unit square. Finite differences are applied on a uniform grid, with $n = 16, 32$, and 64 grid lines in each coordinate direction. We use a V(1,1)-cycle with Gauss–Seidel relaxation, sweeping first over the C-points, then over the F-points. This $C - F$ relaxation is the AMG analogue of a red-black Gauss–Seidel sweep. The results of the three experiments are shown in Table 8.1. Note that we use the standard (unscaled) Euclidean norm because we do not presume any geometric structure. In other words, we do not necessarily have the concept of mesh size. Also, we measure the residual only because the error is not ordinarily available.

In reality, AMG is unnecessary for this problem because it is precisely the case for which the geometric case was designed; nonetheless, the analysis of the performance of AMG on this problem is illuminating. We observe first that AMG exhibits

V-cycle	N	$\|\mathbf{r}\|_2$	Ratio of $\|\mathbf{r}\|_2$	N	$\|\mathbf{r}\|_2$	Ratio of $\|\mathbf{r}\|_2$	N	$\|\mathbf{r}\|_2$	Ratio of $\|\mathbf{r}\|_2$
0	16	$1.73e+01$	–	32	$3.49e+01$	–	64	$7.01e+01$	–
1	16	$6.29e-01$	0.03	32	$2.52e+00$	0.07	64	$6.88e+00$	0.10
2	16	$1.44e-02$	0.02	32	$9.63e-02$	0.03	64	$2.92e-01$	0.04
3	16	$3.56e-04$	0.02	32	$3.84e-03$	0.03	64	$1.28e-02$	0.04
4	16	$9.52e-06$	0.03	32	$1.54e-04$	0.04	64	$5.68e-04$	0.04
5	16	$2.55e-07$	0.03	32	$6.24e-06$	0.04	64	$2.55e-05$	0.04
6	16	$6.87e-09$	0.03	32	$2.52e-07$	0.04	64	$1.16e-06$	0.05
7	16	$1.85e-10$	0.03	32	$1.02e-08$	0.04	64	$5.38e-08$	0.05
8	16	$4.95e-12$	0.03	32	$4.12e-10$	0.04	64	$2.52e-09$	0.05
9	16	$1.37e-13$	0.03	32	$1.66e-11$	0.04	64	$1.19e-10$	0.05
10	16	$5.44e-14$	0.46	32	$7.29e-13$	0.04	64	$6.34e-12$	0.05
11	16	$4.36e-14$	0.80	32	$3.37e-13$	0.46	64	$2.58e-12$	0.41
12				32	$3.17e-13$	0.93	64	$2.69e-12$	1.04

Table 8.1: *The table shows the results of AMG V-cycles applied to boundary value problem (8.17). The 2-norm (Euclidean norm) of the residual after each V-cycle and the ratio of the residual norms on successive V-cycles are tabulated for n = 16, 32, and 64.*

A^{ph}	Number of rows	Number of nonzeros	Density (% full)	Average entries per row
A^h	4096	20224	0.001	4.9
A^{2h}	2048	17922	0.004	8.8
A^{4h}	542	4798	0.016	8.9
A^{8h}	145	1241	0.059	8.6
A^{16h}	38	316	0.219	8.3
A^{32h}	12	90	0.625	7.5
A^{64h}	5	23	0.920	4.6

Table 8.2: *Properties of A^{ph}, $p = 2^k, k = 0, 1, \ldots, 6$, for AMG applied to a two-dimensional problem.*

the same type of convergence that was observed in Chapter 4 with geometric multigrid. The residual norm decreases by a relatively constant factor with each V-cycle. This continues until it levels off after about 12 V-cycles near 10^{-13}, where round-off error is on the order of the residual norm itself. Although we do not show the error, we would find that, as in the geometric case, the level of discretization error is reached after about six V-cycles. The errors would also exhibit the same reduction by a factor of four with each doubling of the resolution, as in the geometric case.

In terms of solver performance, AMG appears to be equivalent to geometric multigrid for this problem. However, there are other factors to be considered in examining AMG. These include the time required to do the setup and the storage required by the method.

We first examine the storage requirements. For the above experiment with $n = 64$, the operators A^{ph}, where $p = 2^k$ on levels $k = 0, 1, \ldots, 6$, have the properties shown in Table 8.2. Several observations are in order here. First, the initial coarse grid, Ω^{2h}, has 2048 points, exactly half the number on the finest grid. This occurs because the five-point Laplacian operator yields the red-black coarsening described earlier (and in Exercise 6). However, each succeeding coarse grid has approximately one-fourth the number of points as the next finer grid. This may be understood by observing that the average number of nonzeros per row, which is 4.9 on the fine grid with the five-point operator (boundary points account for the average being below 5), increases to 8.8 on Ω^{2h} and remains above 8 for the next few grids. Evidently, the Galerkin coarse grid operators have become, effectively, nine-point Laplacian operators! The nine-point Laplacian operator yields full coarsening, in which each succeeding grid has one-fourth the number of points as the next finer grid. It is illuminating (Exercise 9) to examine the interpolation stencils and the formation of the coarse-grid operators to discover why the five-point operator becomes a nine-point operator on the coarse grid.

Summing the number of rows of all operators and dividing by the number of rows in the fine grid matrix (4096) shows that the grid complexity for this problem is 1.68. Thus, we know that storage of the vector of unknowns and the right sides require 1.68 times the space required for the fine-grid quantities. By contrast, the geometric approach has a grid complexity of about $\frac{4}{3}$. The difference can be explained by the fact that the first coarse grid produced by AMG is the red-black grid, while the geometric approach does full coarsening on all grids.

Summing the number of nonzeros in all operators and dividing by the number of nonzeros in the fine-grid operator shows that the operator complexity is 2.205. Thus, the matrices on all levels require just over twice the storage of the original operator A^h. In many cases, operators do not need to be stored in the geometric approach, so there is no explicit operator complexity to be considered. On the other hand, operator complexity also reflects the cost of one relaxation sweep on all grids, so a V(1,1)-cycle of AMG on this problem costs about 4.4 WUs (2.2 WUs on the descent and the ascent). This figure should be compared to $\frac{8}{3}$ WUs in the geometric case (neglecting the cost of intergrid transfers, as before).

Turning to the cost of performing the setup phase, we noted earlier that it is difficult to predict or even account for the amount of arithmetic involved. We can, however, make an observation about the "wall clock time" taken by the setup phase. In the $n = 64$ example, the setup phase required 0.27 seconds, while the solution phase required 0.038 seconds per V-cycle. The setup phase, then, required approximately the same amount of time as seven V-cycles.

These experiments show that AMG performs quite well on a model problem, but that it is more expensive in terms of storage, computation, and time than geometric multigrid. However, AMG is not intended for use on nicely structured problems; it is intended for problems on which geometric multigrid cannot be applied. We close our discussion with such an example.   ⬦⬦

Numerical example. Consider the problem

$$-\nabla \cdot (a(x,y)\nabla u) = f(x,y) \tag{8.18}$$

in a domain Ω with $u = 0$ on $\partial\Omega$. Specifically, let $\Omega = ([0,10] \times [0,5.5])/B$, where B is the disk of radius $\frac{1}{2}$ centered on $(1.25, 2.75)$ shown in Fig. 8.9. The problem is discretized on the unstructured triangulation shown in Fig. 8.10. The discontinuous

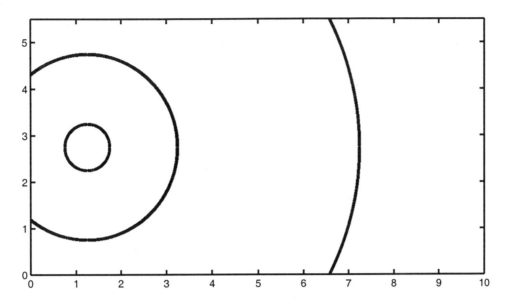

Figure 8.9: *The computational domain for* (8.18).

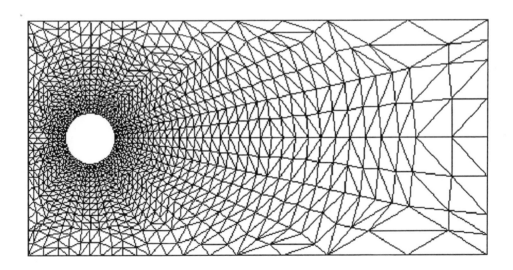

Figure 8.10: *The discretization mesh for* (8.18).

diffusion coefficient is given by

$$a(x,y) = \begin{cases} 0 & \text{if } \sqrt{(x-1.25)^2 + (y-2.75)^2} < \frac{1}{2} \quad \text{(interior of hole)} \\ 1 & \text{if } \frac{1}{2} \le \sqrt{(x-1.25)^2 + (y-2.75)^2} < 2 \\ 1000 & \text{if } 2 \le \sqrt{(x-1.25)^2 + (y-2.75)^2} < 6 \\ 1 & \text{else.} \end{cases}$$

If (x,y) is between the circles, centered about the point $(1.25, 2.75)$, of radius 2

Level	Number of rows	Number of nonzeros	Density (% full)	Average entries per row
0	4192	28832	0.002	6.9
1	2237	29617	0.006	13.2
2	867	14953	0.020	17.2
3	369	7345	0.054	19.9
4	152	3144	0.136	20.7
5	69	1129	0.237	16.4
6	30	322	0.358	10.7
7	20	156	0.390	7.8
8	18	125	0.383	6.9
9	3	9	1.000	3

Table 8.3: *Statistics on coarsening for AMG applied to* (8.18).

V-cycle	$\|\mathbf{r}\|_2$	Ratio of $\|\mathbf{r}\|_2$
0	$1.00e+00$	–
1	$4.84e-02$	0.05
2	$7.59e-03$	0.16
3	$1.72e-03$	0.23
4	$4.60e-04$	0.27
5	$1.32e-04$	0.29
6	$3.95e-05$	0.30
7	$1.19e-05$	0.30
8	$3.66e-06$	0.31
9	$1.12e-06$	0.31
10	$3.43e-07$	0.31
11	$1.05e-07$	0.31
12	$3.22e-08$	0.31

Table 8.4: *Convergence of AMG V-cycles applied to* (8.18) *in terms of the 2-norm (Euclidean norm) of the residual.*

and 6, then the diffusion coefficient is three orders of magnitude larger than it is elsewhere.

This problem has features that would challenge a geometric multigrid algorithm. First, and most important, it is discretized on an unstructured grid. It is not at all obvious how to generate a sequence of coarse grids for this problem. In addition, the discontinuous coefficient causes real difficulty. Since the jump follows circular patterns and is not grid aligned, it is not possible to overcome this problem either by semicoarsening or by line relaxation.

To understand how AMG performs on this problem, we first examine the coarsening statistics in Table 8.3. One interesting phenomenon is that the relative density of the operators (percentage of nonzeros) increases as the grids become coarser. This increasing density can be seen in the last column, where the average number of entries per row increases through several coarse levels. Fortunately, the operator complexity is not adversely affected: a short calculation shows that the operator complexity is 2.97, while the grid complexity is 1.89. It is also interesting to note that the first coarse-grid operator has more actual nonzero coefficients than does the fine-grid operator. This is relatively common for AMG on unstructured grids.

The convergence of the method is summarized in Table 8.4. Using the standard 2-norm, we see that after 11 V-cycles, the residual norm reached 10^{-7} and the process attained an asymptotic convergence factor of 0.31 per V-cycle. Clearly, AMG does not converge as rapidly for this problem as for the previous model problem, where we saw a convergence factor of about 0.1. Nevertheless, in light of the unstructured grid and the strong, non-aligned jump discontinuities in the coefficients, the convergence properties are quite acceptable.   ◇◇

Exercises

1. **Smooth error.** Under the assumptions that $\omega = \|D^{-1/2}AD^{-1/2}\|^{-1}$ and $\|D^{-1/2}AD^{-1/2}\|$ is $O(1)$, show that an implication of the smooth error condition, $\|\left(I - \omega D^{-1}A\right)\mathbf{e}\|_A \approx \|\mathbf{e}\|_A$, is that

   $$(D^{-1}A\mathbf{e}, A\mathbf{e}) \ll (\mathbf{e}, A\mathbf{e}).$$

   Show that this property can be expressed as

   $$\|\mathbf{r}\|_{D^{-1}} \ll \|\mathbf{e}\|_A.$$

2. **Smooth error.** Show that for a symmetric M-matrix A, smooth error varies slowly in the direction of strong influence as follows. Assume $\|r\|_{D^{-1}} \ll \|e\|_A$.

   (a) Show that $\|e\|_A << \|e\|_D$. Hint: $\|e\|_A^2 = (Ae, \mathbf{e}) = (D^{-1/2}Ae, D^{1/2}\mathbf{e}) \le \|r\|_{D^{-1}}\|\mathbf{e}\|_D$.

   (b) Show that if (as is often true) $\sum_{j\neq i}|a_{ij}| \approx a_{ii}$, then

   $$\sum_i \sum_{j\neq i} |a_{ij}|(e_i - e_j)^2 << \sum_i a_{ii}e_i^2.$$

   Hint: Show that $(Ae, \mathbf{e}) = \frac{1}{2}\sum_{ij}(-a_{ij})(e_i - e_j)^2 + \sum_{ij} a_{ij}e_i^2$.

   (c) Conclude that (8.7) must hold on average, that is, for most i.

3. **Deriving interpolation weights.** Carry out the calculation that leads to expression (8.12) for the interpolation weights.

4. **Properties of interpolation weights.** Let A be an M-matrix whose rows sum to zero for interior points, and let the ith row represent an interior point. Show that substituting $e_j = e_i$ in the sum over D_i^s in (8.9) and in the sum over D_i^w in (8.11) produces interpolation coefficients ω_{ij} that are nonnegative and sum to unity.

5. **Enforcing the heuristics.** Construct a simple one-dimensional example in which the two heuristics **H-1** and **H-2** cannot be simultaneously enforced. Hint: Use periodic boundary conditions and odd n.

6. **Coarse-grid selection.** Show that the first pass of the coarsening algorithm (the initial coloring phase) applied to the five-point Laplacian stencil

   $$\frac{1}{h^2}\begin{bmatrix} & -1 & \\ -1 & 4 & -1 \\ & -1 & \end{bmatrix}$$

produces standard red-black coarsening (the C-points are the red squares on a checkerboard; the F-points are the black squares).

7. **Second coloring pass.** The central idea behind the second-pass coloring algorithm is to test each F-point, i, in turn, to ensure that each point in D_i^s depends strongly on at least one point in C_i. When an F-point, i, is found to depend strongly on another F-point, j, that does *not* depend strongly on a point in C_i, then j is made (tentatively) into a C-point; testing of the points in D_i^s then begins again. If all those points now depend strongly on points in C_i, then j is put permanently in C. However, if some other point in D_i^s is found that does not depend strongly on a point in C_i, then i itself is placed in C and j is removed from the tentative C-point list. The process is repeated for the next F-point and continues until all F-points have been treated.

   Apply this second-pass algorithm to the nine-point Laplacian operator with periodic boundaries (Figure 8.5) and determine the final coarsening. Observe that the final coarsening depends on the order in which the F-points are examined.

8. **Using variational properties.** Show that using the variational properties to define interpolation and the coarse-grid operator is optimal in the following way. The two-grid correction scheme, given in the text, corrects the fine-grid approximation $\mathbf{v}^h$ by a coarse-grid interpolant $I_{2h}^h \mathbf{v}^{2h}$ that gives the least error in the sense that

$$\|\mathbf{e}^h - I_{2h}^h \mathbf{v}^{2h}\|_{A^h} = \min_{\mathbf{w}^{2h}} \|\mathbf{e}^h - I_{2h}^h \mathbf{w}^{2h}\|_{A^h},$$

   where $\mathbf{e}^h = \mathbf{u}^h - \mathbf{v}^h$ is the error in $\mathbf{v}^h$.

9. **Coarsening the five-point operator.** Examine the five-point Laplacian operator on a uniform two-dimensional grid, and determine why the Galerkin process alters it to the nine-point operator on the coarse grid. Hint: This can be done symbolically by examining which entries of the stencil will be used to interpolate the various points and then by symbolically carrying out the Galerkin multiplication $A^{2h} = I_h^{2h} A^h I_{2h}^h$.

10. **Smooth error implications.** Show that if $A = D - L - U$ is a symmetric M-matrix and $Q = D + L$ (Gauss–Seidel) or $Q = D$ (Jacobi), then the quantities $(Q^{-1} A \mathbf{e}, \mathbf{r})$ and $(\mathbf{e}, \mathbf{r})$ are nonnegative.

11. **Smooth error.** Show that if $A = D - L - U$ is a symmetric M-matrix and Gauss–Seidel produces errors that satisfy

$$\left\|\left(I - (D + L)^{-1} A\right) \mathbf{e}\right\|_A \approx \|\mathbf{e}\|_A,$$

   then

$$\|\mathbf{r}\|_{D^{-1}} \ll \|\mathbf{e}\|_A.$$

   Hint: Show first that $\|(D + L)^{-1} A \mathbf{e}\|_A \ll \|e\|_A$. Then show and use the fact that $(D + L^T)^{-1} A (D + L)^{-1} \leq D^{-1}$.

12. **V-cycle costs.** Let the grid complexity be denoted σ^Ω and the operator complexity be denoted σ^A. In addition, define the following quantities: κ^A, the average number of nonzero entries per row over all levels; κ^I, the average

number of interpolation points per F-point; n_m^A, the number of nonzero entries in A^m; and n_m^C and n_m^F, the number of C- and F-points, respectively, on grid Ω^m.

(a) Show that the number of floating-point operations on level m for one relaxation sweep, residual transfer, and interpolation are $2n_m^A$, $2n_m^A + 2\kappa^I n_m^F$, and $n_m^C + 2\kappa^I n_m^F$, respectively.

(b) Noting that

$$\sum_m n_m^F \approx n,$$

show that the total flop count for a $V(\nu_1, \nu_2)$-cycle is given approximately by

$$n(2(\nu + 1)\kappa^A \sigma^\Omega + 4\kappa^I + \sigma^\Omega - 1),$$

where $\nu = \nu_1 + \nu_2$.

Chapter 9

Multilevel Adaptive Methods

Numerical problems often exhibit special features in small local regions that require resolution and accuracy well beyond what is required in the rest of the domain. In numerical weather modeling, isolated phenomena such as tornados or storm fronts may demand substantially enhanced accuracy. The numerical simulation of the flight of an aircraft may require especially accurate approximations around the fuselage or wings. In these and many other cases, it is wasteful to let local accuracy requirements dictate the global discretization and solution process. The goal of this chapter is to understand how to treat local demands in a multilevel context without overburdening the overall computation.

There is a wide variety of methods that effectively treat local demands. Here we consider the so-called *fast adaptive composite* grid method (FAC) [14]. Its distinctive features are that it always works with uniform grids and subgrids, and it is in tune with the variational theme that appears throughout this book. We begin with a one-dimensional example.

Consider the two-point boundary value problem

$$
\begin{aligned}
-u''(x) &= f(x), \quad 0 < x < 1, \\
u(0) = u(1) &= 0.
\end{aligned}
$$

To keep matters simple, assume that a *local fine grid* with mesh size $h = \frac{1}{8}$ is needed on the interval $(\frac{1}{2}, 1)$ to resolve some special feature near $x = \frac{3}{4}$, but that a mesh size of $2h = \frac{1}{4}$ is deemed adequate for the rest of the domain. This need might arise, for example, in the presence of a source term f that is nonzero only near $x = \frac{3}{4}$ (Fig. 9.1).

Imagine first that we solve the problem on a *global* grid, Ω^h, with mesh size $h = \frac{1}{8}$, using a two-grid method based on linear interpolation, full weighting, and the variational properties as described in Chapter 5. We write the discrete problem as

$$ A^h \mathbf{u}^h = \mathbf{f}^h. \tag{9.1} $$

In component form, we have

$$
\begin{aligned}
\frac{-u_{i-1}^h + 2u_i^h - u_{i+1}^h}{h^2} &= f_i^h, \quad 1 \le i \le n, \\
u_0^h = u_{n+1}^h &= 0,
\end{aligned}
$$

where $n = 7$ for this case.

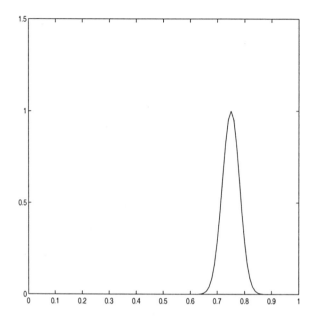

Figure 9.1: *Graph of a sharply peaked impulse function centered at $x = \frac{3}{4}$.*

This simple uniform grid involves unnecessary computation: the accuracy required *locally* at $x = \frac{3}{4}$ dictates the *global* mesh size $h = \frac{1}{8}$, forcing us to solve the problem with too many grid points. To reduce this waste, we can start by eliminating relaxation (the most expensive multigrid process) at fine-grid points where it is not needed, that is, in $[0, \frac{1}{2}]$. However, we can do better. As we will see, it is possible to reduce costs further by selectively eliminating interpolation, restriction, and the computation of residuals.

Consider Fig. 9.2, which shows the fine grid Ω^h with $n = 7$ interior points and the underlying coarse grid Ω^{2h} with $\frac{n-1}{2} = 3$ interior points. Regular multigrid relaxes on all fine-grid points, denoted by $*, \bullet$, and $\circ$, then performs coarse-grid correction based on the points denoted by $\times$.

With this notation, let us see how we might eliminate points of Ω^h in $[0, \frac{1}{2})$. Restricting relaxation to the local fine grid in $(\frac{1}{2}, 1)$ means that the Ω^h residuals change at $*$ and $\bullet$ points, because the approximation $\mathbf{v}^h$ changes at $*$ points. However, *fine-grid residuals do not change at $\circ$ points between successive coarse-grid solves*. Similarly, the coarse-grid residual does not change at $\times = \frac{1}{4}$ because it is obtained by restriction from $\circ$ points.

For the moment, suppose that we continue to store $\mathbf{v}^h$ at $\circ$ points, just for convenience. We must be sure that the algorithm starts with the correct right sides, f_i^{2h}, on the coarse grid, particularly at the point $\times = \frac{1}{4}$. We begin with a zero initial guess, $\mathbf{v}^h = \mathbf{0}$, so the residual at the coarse-grid point $\times = \frac{1}{4}$ is just the restriction of the fine-grid source term f_i^h. With this observation, we can give our first scheme:

- Initialize $\mathbf{v}^h = \mathbf{0}$ and $f_1^{2h} \equiv \left(I_h^{2h} \mathbf{f}^h\right)_1 = \frac{f_1^h + 2f_2^h + f_3^h}{4}$.

- Relax on $\mathbf{v}^h$ on the local fine grid $\{x_5^h, x_6^h, x_7^h\} = \{\frac{5}{8}, \frac{3}{4}, \frac{7}{8}\}$.

Ω^h	x_0^h	x_1^h	x_2^h	x_3^h	x_4^h	x_5^h	x_6^h	x_7^h	x_8^h
Ω^h	$\odot$	$\circ$	$\circ$	$\circ$	$\bullet$	$*$	$*$	$*$	$\odot$
$x =$	0	$\frac{1}{8}$	$\frac{1}{4}$	$\frac{3}{8}$	$\frac{1}{2}$	$\frac{5}{8}$	$\frac{3}{4}$	$\frac{7}{8}$	1
Ω^{2h}	$\odot$		$\times$		$\times$		$\times$		$\odot$
Ω^{2h}	x_0^{2h}		x_1^{2h}		x_2^{2h}		x_3^{2h}		x_4^{2h}

Figure 9.2: *Local fine-grid points on Ω^h are denoted by $*$ and the interface point is denoted by ($\bullet$). Computation at the remaining fine-grid points ($\circ$) is eliminated. The coarse-grid points ($\times$) comprise Ω^{2h}, while $\odot$ represents boundary points.*

- Compute fine-grid residual $\mathbf{r}^h = \mathbf{f}^h - A^h\mathbf{v}^h$ and transfer it to the coarse grid:

$$f_2^{2h} \leftarrow \frac{r_3^h + 2r_4^h + r_5^h}{4}, \; f_3^{2h} \leftarrow \frac{r_5^h + 2r_6^h + r_7^h}{4}.$$

- Compute an approximation $\mathbf{v}^{2h}$ to the coarse-grid residual equation $A^{2h}\mathbf{u}^{2h} = \mathbf{f}^{2h}$.

- Update the residual at $\times = \frac{1}{4}$ for use in later cycles: $f_1^{2h} \leftarrow f_1^{2h} - \frac{-v_0^{2h} + 2v_1^{2h} - v_2^{2h}}{(2h)^2}$.

- Interpolate the correction and update the approximation: $\mathbf{v}^h \leftarrow \mathbf{v}^h + I_{2h}^h\mathbf{v}^{2h}$.

At this point, reducing the number of computations further becomes a little tricky. It might seem at first glance that the fine-grid approximation, $\mathbf{v}^h$, is never used outside of $[\frac{1}{2}, 1]$, but this is not quite true: the residual transfer in the third step involves the residuals $\mathbf{r}^h$ at $\bullet = \frac{1}{2}$ and $\circ = \frac{3}{8}$, which in turn involve $\mathbf{v}^h$ at $\circ = \frac{1}{4}$ and $\circ = \frac{3}{8}$. Thus, we need to store $\mathbf{v}^h$ at these two *border points* just outside the local fine grid. On the other hand, we do not need to store v_1^h at $\circ = \frac{1}{8}$. This observation suggests introducing a new variable w_1^{2h} that accumulates the coarse-grid solution at $\times = \frac{1}{4}$, which underlies the fine-grid point $\circ = \frac{1}{4}$. We can identify $w_1^{2h} = v_2^h$, allowing us to keep track of the solution at this border point. The solution at the other border point, $\circ = \frac{3}{8}$, will be found by interpolation. The improved algorithm now appears as follows:

- Initialize $\mathbf{v}^h = 0, w_1^{2h} = 0$, and $f_1^{2h} \equiv \left(I_h^{2h}\mathbf{f}^h\right)_1 = \frac{f_1^h + 2f_2^h + f_3^h}{4}$.

- Relax on $\mathbf{v}^h$ on the local fine grid $\{x_5^h, x_6^h, x_7^h\} = \{\frac{5}{8}, \frac{3}{4}, \frac{7}{8}\}$.

- Compute fine-grid residual $\mathbf{r}^h = \mathbf{f}^h - A^h\mathbf{v}^h$ on the local fine grid, its interface point $x_4^h = \frac{1}{2}$, and the border point $x_3^h = \frac{3}{8}$, and transfer it to the coarse grid:

$$f_2^{2h} \leftarrow \frac{r_3^h + 2r_4^h + r_5^h}{4}, \; f_3^{2h} \leftarrow \frac{r_5^h + 2r_6^h + r_7^h}{4}.$$

- Compute an approximation $\mathbf{v}^{2h}$ to the coarse-grid residual equation $A^{2h}\mathbf{u}^{2h} = \mathbf{f}^{2h}$.

- Update the residual at $\times = \frac{1}{4}$ for use in later cycles: $f_1^{2h} \leftarrow f_1^{2h} - \frac{-v_0^{2h}+2v_1^{2h}-v_2^{2h}}{(2h)^2}$.

- Update the coarse-grid approximation at $\times = \frac{1}{4}$: $w_1^{2h} \leftarrow w_1^{2h} + v_1^{2h}$.

- Interpolate the correction and update the approximation: $\mathbf{v}^h \leftarrow \mathbf{v}^h + I_{2h}^h \mathbf{v}^{2h}$ everywhere except at $\circ = \frac{1}{8}, \frac{1}{4}$.

Note that we do not interpolate to $\circ = \frac{1}{4}$, because the solution at this point is held in w_1^{2h}, nor to $\circ = \frac{1}{8}$, because a fine-grid solution is not needed at this point.

We have devoted an absurd amount of effort here to avoiding computation at the single point $\circ = \frac{1}{8}$. However, it is important to imagine situations in which the local fine grid comprises a very small fraction of the domain. In such situations, points like $\circ = \frac{1}{8}$ predominate and the savings can be significant. As we see shortly, the savings can be even greater for two-dimensional problems.

We can take one more step and effectively remove the border points from the computation. Consider the residual f_2^{2h} at the interface point $\bullet = \frac{1}{2}$; it is obtained by full weighting the fine-grid residuals r_3^h, r_4^h, and r_5^h. Because w_1^{2h} accumulates the solution at $\circ = \frac{1}{4}$, we have $v_2^h = w_1^{2h}$; and because the solution at $\circ = \frac{3}{8}$ is determined by interpolation, we have $v_3^h = (w_1^{2h} + v_4^h)/2$. Thus, the fine-grid residuals that contribute to f_2^{2h} are computed as follows:

$$
\begin{aligned}
r_3^h &= f_3^h - \frac{-v_2^h + 2v_3^h - v_4^h}{h^2} \\
&= f_3^h - \frac{-w_1^{2h} + w_1^{2h} + v_4^h - v_4^h}{h^2} \\
&= f_3^h, \qquad\qquad\qquad\qquad\qquad\qquad\qquad (9.2) \\
r_4^h &= f_4^h - \frac{-(w_1^{2h} + v_4^h)/2 + 2v_4^h - v_5^h}{h^2} \\
&= f_4^h - \frac{-\frac{1}{2}w_1^{2h} + \frac{3}{2}v_4^h - v_5^h}{h^2}, \qquad\qquad\quad (9.3) \\
r_5^h &= f_5^h - \frac{-v_4^h + 2v_5^h - v_6^h}{h^2} . \qquad\qquad\qquad (9.4)
\end{aligned}
$$

Writing $g_2^{2h} \equiv \frac{1}{4}\left(f_3^h + 2f_4^h + f_5^h\right)$, a bit of algebra yields the residual f_2^{2h} at the interface point:

$$
\begin{aligned}
f_2^{2h} &\equiv \frac{1}{4}\left(r_3^h + 2r_4^h + r_5^h\right) \qquad\qquad\qquad\quad (9.5) \\
&= g_2^{2h} - \frac{-w_1^{2h} + 2v_4^h - v_6^h}{(2h)^2} .
\end{aligned}
$$

Note that the stencil on the right side of (9.5) corresponds to the usual Ω^{2h} stencil for the residual. This is a result of the variational properties at work. Yet, even in more general cases, it is possible to compute the *result* of transferring the fine-grid residual to the interface point(s). In turn, this allows us to eliminate the border points. These simplifications lead to our final algorithm.

Fast Adaptive Composite Grid Method (FAC)

- Initialize $\mathbf{v}^h = \mathbf{0}, w_1^{2h} = 0$, and $f_1^{2h} \equiv \left(I_h^{2h}\mathbf{f}^h\right)_1 = \frac{f_1^h + 2f_2^h + f_3^h}{4}$.

- Relax on $\mathbf{v}^h$ on the local fine grid $\{x_5^h, x_6^h, x_7^h\} = \{\frac{5}{8}, \frac{3}{4}, \frac{7}{8}\}$.

- Compute the right sides for the local coarse grid: $f_2^{2h} \leftarrow g_2^{2h} - \frac{-w_1^{2h} + 2v_4^h - v_6^h}{(2h)^2}$,
  $f_3^{2h} \leftarrow \frac{r_5^h + 2r_6^h + r_7^h}{4}$.

- Compute an approximation $\mathbf{v}^{2h}$ to the coarse-grid residual equation $A^{2h}\mathbf{u}^{2h} = \mathbf{f}^{2h}$.

- Update the residual at $\times = \frac{1}{4}$ for use in later cycles: $f_1^{2h} \leftarrow f_1^{2h} - \frac{-v_0^{2h} + 2v_1^{2h} - v_2^{2h}}{(2h)^2}$.

- Update the coarse-grid approximation at $\times = \frac{1}{4}$: $w_1^{2h} \leftarrow w_1^{2h} + v_1^{2h}$.

- Interpolate the correction and update the approximation: $\mathbf{v}^h \leftarrow \mathbf{v}^h + I_{2h}^h \mathbf{v}^{2h}$
  at the local fine grid and the interface points $\{\frac{1}{2}, \frac{5}{8}, \frac{3}{4}, \frac{7}{8}\}$.

Summary of FAC Terms

- The *local fine grid* is the finest grid used for computation; it covers only that part of the domain where additional resolution is needed. FAC avoids computation on a global fine grid.

- The *global coarse grid*, Ω^{2h}, is the finest grid used for computation that covers the entire domain. The notation suggests a mesh refinement factor of two, but larger factors are permitted.

- *Interface points* are the boundary points of the local fine grid.

- *Border points* are fine-grid points that lie outside the local fine grid. They are used temporarily to develop special interface stencils.

- *Boundary points* are the usual points of the domain where boundary conditions apply.

- *Slave points*, which appear only in two and higher dimensions, are interface points that do not correspond to coarse-grid points.

- The *composite grid* is the combination of the fine- and coarse-grid points on which the discrete solution is ultimately determined.

It is important to see FAC from a different perspective. We developed the scheme by eliminating the ∘ points from the fine grid and producing the top two uniform grids (Ω^h and Ω^{2h}) shown in Fig. 9.3. The question we need to ask now is: At what points are we actually approximating the solution? When we use FAC, the solution is approximated at the *composite grid* (Ω^c in Fig. 9.3) consisting of

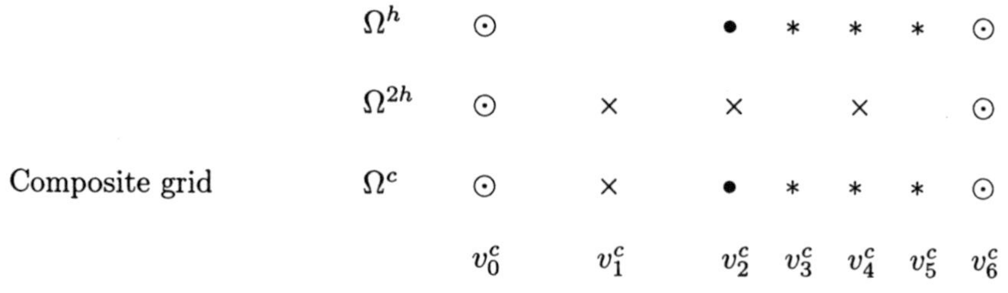

$$
\begin{array}{cccccccc}
\Omega^h & \odot & & & \bullet & * & * & * & \odot \\
\Omega^{2h} & \odot & \times & & \times & & \times & & \odot \\
\Omega^c & \odot & \times & & \bullet & * & * & * & \odot \\
& v_0^c & v_1^c & & v_2^c & v_3^c & v_4^c & v_5^c & v_6^c
\end{array}
$$

Composite grid

Figure 9.3: *FAC is done entirely on uniform grids, such as the top two grids shown above. However, the solution is in effect determined on the nonuniform composite grid shown at the bottom.*

- local fine-grid points ($*$),

- interface points ($\bullet$), and

- coarse-grid points that do not lie under the local fine-grid or interface points ($\times$ points in $(0, \frac{1}{2})$).

FAC operates only on uniform grids, but it effectively solves problems on compositive grids. It is fully consistent with multigrid principles because, in the variational case, FAC is equivalent to solving the problem by multigrid using global grids, but local relaxation. This property distinguishes FAC from other adaptive methods.

While we never need to construct the composite grid, it serves as an important conceptual tool for understanding the refinement algorithm that we just developed. For example, because FAC is a method for solving the problem on the composite grid, we should be able to derive the associated composite-grid equations. This will allow us to see what algebraic problem FAC actually solves.

To derive these composite-grid equations, first consider the composite-grid approximation $\mathbf{v}^c$ that FAC produces (we include the boundary points v_0^c and v_6^c for completeness):

$$
v_0^c = 0, \; v_1^c = w_1^{2h}, \; v_2^c = v_4^h, \; v_3^c = v_5^h, \; v_4^c = v_6^h, \; v_5^c = v_7^h, \; v_6^c = 0.
$$

We also need to define the source vector $\mathbf{f}^c$. Leaving the definition of f_2^c aside for the moment, we have

$$
f_1^c = \frac{f_1^h + 2f_2^h + f_3^h}{4}, \; f_3^c = f_5^h, \; f_4^c = f_6^h, \; f_5^c = f_7^h.
$$

Now imagine that the FAC scheme has converged, by which we mean that $\mathbf{v}^c$ does not change from one cycle to the next. Inspecting the algorithm shows that the stencils in the uniform regions ($\times$ and $\circ$ points) are the usual ones with the appropriate mesh sizes:

$$
\frac{-v_0^c + 2v_1^c - v_2^c}{(2h)^2} = f_1^c,
$$
$$
\frac{-v_{i-1}^c + 2v_i^c - v_{i+1}^c}{h^2} = f_i^c, \quad 3 \le i \le 5. \tag{9.6}
$$

The difficult stencil is the one at the interface point $\bullet = \frac{1}{2}$. Consider (9.5) and remember we are assuming that FAC has converged, so relaxation does not change v_5^h. This means that r_5^h must be zero. Using this observation and substituting (9.2) and (9.3) into (9.5) gives us

$$
\begin{aligned}
f_2^{2h} &= \frac{1}{4}\left(r_3^h + 2r_4^h\right) \\
&= \frac{1}{4}\left(f_3^h + 2f_4^h\right) - \left(-w_1^{2h} + 3v_4^h - 2v_5^h\right)/(2h)^2 .
\end{aligned}
$$

We must have $f_2^{2h} = 0$, or else the FAC coarse-grid step would compute a nonzero correction to $\mathbf{v}^c$ and we could not have converged. We thus conclude that the stencil at $\bullet = \frac{1}{2}$ is

$$
\frac{-w_1^{2h} + 3v_4^h - 2v_5^h}{(2h)^2} = \frac{1}{4}\left(f_3^h + 2f_4^h\right) .
$$

In composite-grid terms, with $f_2^c \equiv \frac{1}{4}\left(f_3^h + 2f_4^h\right)$, we have

$$
\frac{-v_1^c + 3v_2^c - 2v_3^c}{(2h)^2} = f_2^c. \tag{9.7}
$$

One of the main concerns in the design of any numerical algorithm is how to measure the error. This is a complex issue that involves various error sources (discretization error, algebraic error, and floating-point errors), the choice of norms, and various ways to approximate the error. Most of these issue are beyond the scope of this book, but a few brief comments are in order:

- Estimating discretization error is of particular interest in adaptive methods because it can be used to decide where to refine. There are many conventional approaches that involve solving local problems or estimating higher derivatives of the emerging solution. However, the presence of several levels of discretization enables multilevel methods to use extrapolation: you can compare the approximations on two or more levels to predict the error on the finest level. For this purpose, it is convenient to have the full approximations available on each level, so it can be useful to rewrite the FAC *correction* scheme that we have presented as an FAS scheme (Exercise 1).

- The algebraic error can also be estimated in several ways, but a natural estimate is to apply a norm to the residual of the composite-grid equations. Remember, the composite grid consists of the uniform fine grid in the refinement region, so any point within this region has a standard fine-grid equation. Similarly, points outside the refinement region have standard coarse-grid equations. But the interface has special equations (9.7) that require special residuals. Putting this together for our example gives

$$
\begin{aligned}
r_1^c &= \frac{1}{4}\left(f_1^h + 2f_2^h + f_3^h\right) - \frac{2w_1^{2h} - v_4^h}{(2h)^2}, &&\text{standard coarse-grid residual;} \\
r_2^c &= \frac{1}{4}\left(f_3^h + 2f_4^h\right) - \frac{-w_1^{2h} + 3v_4^h - 2v_5^h}{(2h)^2}, &&\text{interface residual from (9.7);} \\
r_i^c &= f_i^h - \frac{-v_{i+1}^h + 2v_{i+2}^h - v_{i+3}^h}{h^2}, \quad 3 \le i \le 5, &&\text{standard fine-grid residual.}
\end{aligned}
$$

- Scaling the error norms is particularly important in adaptive methods because of the presence of different levels of resolution. The proper scales depend on the discretization and other factors. For our example, you can think of the residual $\mathbf{r}^c$ as a source term that is constant in a neighborhood of each node. Thus, r_1^c represents the value in the interval $(\frac{1}{4} - h, \frac{1}{4} + h)$, r_2^c represents the value in the interval $(\frac{1}{2} - h, \frac{1}{2} + \frac{h}{2})$, and r_i^c represents the value in the interval $(\frac{i+2}{8} - \frac{h}{2}, \frac{i+2}{8} + \frac{h}{2})$ for $i = 3, 4, 5$. (This does not cover the interval $[0, 1]$, but we can simply assume that the source is zero in $[0, \frac{1}{16})$ and $(\frac{15}{16}, 1]$.) Using the premise that the residual represents a piecewise constant source term, we can obtain the norm of $\mathbf{r}^c$ by taking the integral of the square of the source term defined on $(0, 1)$, which yields

$$\|\mathbf{r}^c\| = \sqrt{2h(r_1^c)^2 + \frac{3}{2}h(r_2^c)^2 + h\sum_3^5 (r_i^c)^2}\,.$$

Two-Dimensional Problems

With some modification, the above ideas can be extended to two-dimensional problems. Figure 9.4 shows a square domain in the plane with a local fine grid in the northeast corner. There are many points in this two-dimensional grid that are analogous to $\circ = \frac{1}{8}$, at which fine-grid computations can be neglected. For this reason,

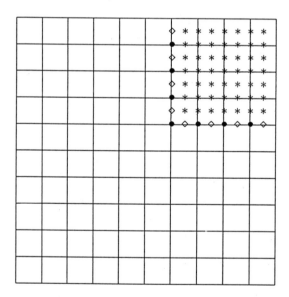

Figure 9.4: *Composite grid on a two-dimensional domain. The coarse-grid points are at the intersection of the solid horizontal and vertical lines. The local fine grid consists of $*$ points. Interface points coinciding with coarse-grid points are marked by $\bullet$. The new slave points ($\diamond$) are interface points that do not overlay a fine-grid point. Boundary points have been omitted.*

the FAC scheme shows substantial savings over standard multigrid on a global fine grid.

A new type of point arises in two-dimensional problems. Denoted by $\diamond$ in Fig. 9.4, the *slave points* are interface points that do not correspond to a coarse-grid point. These points are needed to provide a complete set of boundary points for the local fine grid. They are determined simply by averaging the two neighboring interface point values. For example, in Fig. 9.4, a typical slave point on a horizontal interface would be determined by

$$v_{i+1,j}^h = \frac{v_{i,j}^h + v_{i+2,j}^h}{2}.$$

The one-dimensional FAC method can be generalized in all respects to two dimensions. Special restriction formulas at the interface points corresponding to (9.5) can be used to eliminate border points. Similarly, with some additional work, it is possible to find the coarse-grid and interface equations as was done in one dimension (Exercise 4). However, our goal is not to expose these details (which can be found in [14]). Rather, we will only outline the flow of the two-dimensional FAC method in fairly qualitative terms.

Fast Adaptive Composite Grid Method (FAC) in Two Dimensions

- Initialize $\mathbf{v}^h = 0$. Initialize $\mathbf{w}^{2h} = 0$ and $\mathbf{f}^{2h} = I_h^{2h}\mathbf{f}^h$ at coarse-grid points outside the local fine grid.

- Relax on $\mathbf{v}^h$ on the local fine grid.

- Compute the right sides for the local coarse grid using special stencils at the interface points and the usual residual transfer $\mathbf{f}^{2h} = I_h^{2h}\mathbf{f}^h$ at points underlying the local fine grid.

- Compute an approximation $\mathbf{v}^{2h}$ to the coarse-grid residual equation $A^{2h}\mathbf{u}^{2h} = \mathbf{f}^{2h}$.

- Update the coarse-grid residual for use in later cycles: $\mathbf{f}^{2h} \leftarrow \mathbf{f}^{2h} - A^{2h}\mathbf{v}^{2h}$ at points outside the local fine grid.

- Update the coarse-grid approximation: $\mathbf{w}^{2h} \leftarrow \mathbf{w}^{2h} + \mathbf{v}^{2h}$ at coarse-grid points outside the local fine grid.

- Interpolate the correction and update the approximation: $\mathbf{v}^h \leftarrow \mathbf{v}^h + I_{2h}^h\mathbf{v}^{2h}$ at the local fine grid and interface points. Interpolate to the slave points from their interface neighbors.

We have presented FAC as a two-grid scheme. Like other multigrid methods, it is done recursively in practice: in the fourth step of the above algorithm, the solution of the global coarse-grid problem is done using further applications of the basic two-grid scheme. This procedure leads to V-cycles defined on a sequence of grids, possibly involving several global coarse grids together with several telescoping finer grids.

An FMG–FAC scheme is more subtle. One could apply a straightforward FMG cycling scheme to the uniform grids that are used in the V-cycle process. However, this would be too costly when there are many levels with few points per level (Exercise 3). A more efficient approach is to use a sequence of increasingly coarse composite grids as the basis for FMG, a strategy that takes us beyond the scope of this book.

The fundamental basis for FAC is the composite grid. In fact, probably the most effective way to apply FAC to a given problem is to *start* by developing accurate composite-grid equations. The FAC algorithm is then a natural extension of multigrid, where the finest level is the composite grid, relaxation is simply restricted to the local fine grid, and coarsening is based on restricting the composite grid residual to the global coarse grid.

A popular alternative to FAC is the *multilevel adaptive technique*, or *MLAT* [3]. Although these two approaches can lead to similar algorithms in the end, the basis for their development is very different. While FAC is based on the composite grid, MLAT comes from generalizing the full approximation scheme (FAS). A little inspection of FAS, as developed in Chapter 6, shows that the finest grid need not be global: one can relax only on a local fine grid and correct the coarse-grid equations by FAS only at points that lie under it. MLAT can thus be developed from a global-grid FAS code, and it often proves effective for handling adaptive refinement.

Numerical example. To illustrate FAC performance in a simple setting, FAC is applied to Poisson's equation with homogeneous Dirichlet boundary conditions on the unit square in the plane. The exact two-spike solution, as shown in Fig. 9.5, has values of magnitude 1 but opposite sign at $(\frac{1}{4}, \frac{1}{4})$ and $(\frac{3}{4}, \frac{3}{4})$, falls off steeply in the neighborhood of these points, and is zero at the boundary. In the tests, we used composite grids constructed from a global grid of mesh size h and refinement

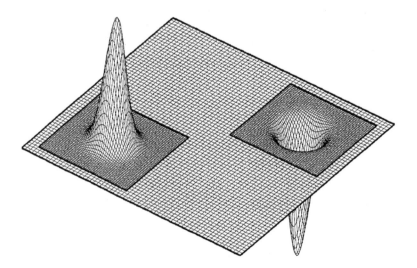

Figure 9.5: *Two-spike solution of the composite grid problem based on a* 63 × 63 *interior global grid, with mesh size* $h = \frac{1}{64}$, *and* 51×51 *patches of mesh size* $\frac{h}{2} = \frac{1}{128}$ *centered at* $(\frac{1}{4}, \frac{1}{4})$ *and* $(\frac{3}{4}, \frac{3}{4})$.

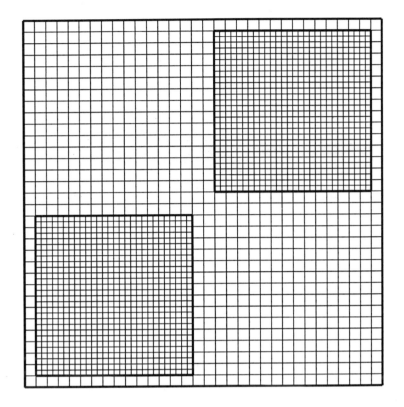

Figure 9.6: *Aerial view of a composite grid based on a 31×31 interior global grid of mesh size $h = \frac{1}{32}$ and 27×27 patches of mesh size $\frac{h}{2} = \frac{1}{64}$ centered at $\left(\frac{1}{4}, \frac{1}{4}\right)$ and $\left(\frac{3}{4}, \frac{3}{4}\right)$.*

patches of size $\frac{h}{2}$ in the subregions $\left[\frac{1}{32}, \frac{15}{32}\right] \times \left[\frac{1}{32}, \frac{15}{32}\right]$ and $\left[\frac{17}{32}, \frac{31}{32}\right] \times \left[\frac{17}{32}, \frac{31}{32}\right]$. Tests were done for $h = \frac{1}{32}, \frac{1}{64}, \frac{1}{128}$, with the case $h = \frac{1}{64}$ shown in Fig. 9.6.

Table 9.1 contains the results of applying a V(1,0)-cycle to each of the three composite grids. The convergence factor was obtained by applying many cycles to the homogeneous problem (with zero right side), starting with a random initial guess, and observing the worst-case composite-grid residual reduction factor. Note

Global h	Convergence factor	Discrete L^2 norm of discretization error
1/32	0.362	2.34e − 2
1/64	0.367	5.742 − 3
1/128	0.365	1.43e − 3

Table 9.1: *Estimates for asymptotic V(1,0)-cycle convergence factors and discrete L^2 norm of the discretization errors. FAC was applied using composite grids with three different global grids of mesh size h, each with two local patches of mesh size $\frac{h}{2}$.*

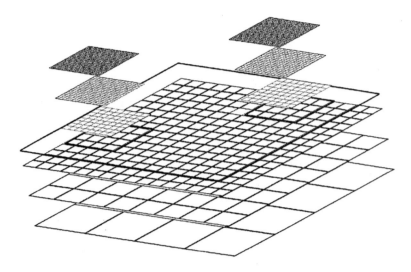

Figure 9.7: *Side view of a global 15×15 grid with two coarser supporting grids and three successively finer local grids.*

the apparent h-independence of these convergence factors. These factors are quite acceptable considering the fact that V(1,0)-cycles were used. Also shown are the discretization error estimates, obtained by applying several V(1,0)-cycles to the two-spike problem and measuring the difference, in the discrete L^2 norm, between the computed and the exact solutions, evaluated on the composite grid. Note the apparent $O(h^2)$ behavior of the discretization error.

Another illustration of the grids that comprise the composite grids is given in Fig. 9.7. Here we show a side view of a global grid of mesh size $h = \frac{1}{8}$ and local grids of mesh size $\frac{h}{8} = \frac{1}{64}$ in the subregions $[\frac{1}{8}, \frac{3}{8}] \times [\frac{1}{8}, \frac{3}{8}]$ and $[\frac{5}{8}, \frac{7}{8}] \times [\frac{5}{8}, \frac{7}{8}]$. Note that the effective refinement factor (ratio of mesh sizes of the finest patch and finest global grid) is eight. Large refinement factors are permissible with FAC because local grids with factors of two are included between the global and finest local grids. For similar reasons, the finest global grid is supported by a full sequence of coarser global grids, as shown.   ◇◇

Exercises

1. **FAS version of FAC.** The FAC algorithm we have described is a correction scheme in which the coarse-grid problem is the residual equation. Rewrite FAC for the one-dimensional model problem as an FAS scheme in which the full approximation equations are solved on the coarse grid.

2. **Higher ⟩ ʒh refinement factors.** We developed the FAC scheme assuming a *mesh refinement factor* of two: the global coarse-grid mesh size is $2h$, while the local fine-grid mesh size is h. Rewrite FAC for solving the one-dimensional model problem when the coarse-grid mesh size is $4h = \frac{1}{4}$ and the fine-grid mesh size is $h = \frac{1}{16}$. Assume that the local fine grid covers $(\frac{1}{2}, 1)$ as

before. (For full efficiency, FAC should include the intermediate grid of mesh size $2h$ that covers $(\frac{1}{2}, 1)$, but the two-grid scheme suffices here.)

3. **Cycling cost.** Show that V-cycles maintain $O(n)$ complexity for locally refined grids in the sense that the cost of an FAC V-cycle is proportional to the number, n, of *composite-grid* points. For concreteness, consider a sequence of q grids on the unit square in the plane, constructed by starting from a global $m \times m$ grid and refining successive grids by a factor of two in the northeast quadrant (so that each grid is $m \times m$). Now show that W-cycles do *not* maintain $O(n)$ complexity. Similarly, show that FMG based on these uniform grids also does not maintain $O(n)$ complexity. Finally, show that FMG based on coarsened *composite* grids does maintain $O(n)$ complexity. (Coarse composite grids are formed by coarsening all grids in the sequence, so the first coarse composite grid is formed from q grids of size $\frac{m}{2} \times \frac{m}{2}$.)

4. **Two-dimensional interface source terms.** Deriving the Ω^{2h} and Ω^c interface equations is somewhat complicated in two dimensions. Take the first step in this direction by forming the source term for Ω^{2h} and Ω^c at the interface corner point in Fig. 9.4.

5. **Symmetry.** Show that the composite-grid matrix A^c is nonsymmetric by showing that the coefficient in (9.6) that connects v_2^c to v_3^c is not equal to the corresponding coefficient in (9.7). Show, however, that symmetry can be retained by rescaling: the matrix $D^c A^c$ is symmetric, where D^c is the 5×5 diagonal matrix with entries $d_1^c = d_2^c = 2h$ and $d_3^c = d_4^c = d_5^c = h$.

6. **Neumann problem.** Rederive FAC applied to Neumann problem (7.1) of Chapter 7. Show that the principles of Neuman boundary conditions apply by showing that the null space and range of the rescaled composite-grid matrix, $D^c A^c$ of Exercise 5, are equal and consist of scalar multiples of the constant vector **1**.

Chapter 10

Finite Elements

Until now, our development of multigrid methods has relied on finite differences. However, there are several other discretization methods, all of which have their place in computation. We now turn to the finite element method. This approach may seem more abstract and less direct at the outset, but its use of variational properties provides a powerful framework that is well suited to developing and analyzing multigrid methods. It is impossible to cover finite elements completely in this short chapter, so we focus instead on surveying the critical ideas, leaving the important details to the exercises and further study.

Said simply, finite difference methods replace the problem domain by a grid and produce a vector whose components are approximations to the solution at the grid points. On the other hand, finite element methods partition the problem domain into subregions and produce a simple function in each subregion that approximates the solution. Finite element methods can be applied to two major categories of problems: differential equations and functional minimization. When the differential equation is defined in terms of a self-adjoint linear operator and the functional is defined appropriately, these two categories coincide. To exploit this fortunate duality between differential equations and functional minimization, we focus on the self-adjoint case.

Let Ω be a suitably nice bounded domain in the plane with boundary $\partial\Omega$. Consider the following Poisson equation with homogeneous Dirichlet boundary conditions:

$$
\begin{aligned}
Lu \equiv -u_{xx} - u_{yy} &= f &&\text{in } \Omega, \\
u &= 0 &&\text{on } \partial\Omega.
\end{aligned}
$$

We first need to introduce some new tools and notation. Most of our time will be spent in the space $L^2(\Omega)$ of square integrable functions on Ω (functions that satisfy $\int_\Omega u^2 \, d\Omega < \infty$) equipped with the $L^2(\Omega)$ inner product $(\cdot, \cdot)$ defined by

$$
(u, v) = \int_\Omega uv \, d\Omega.
$$

Because we will be taking partial derivatives, it is not enough for the functions we consider just to be $L^2(\Omega)$ integrable. We need to know that they are differentiable. We will be vague about this for the moment and just assume that the functions reside in a subspace $\mathcal{H} \subset L^2(\Omega)$ that contains *suitably differentiable* functions. We also assume that the functions in $\mathcal{H}$ vanish on $\partial\Omega$ so that they satisfy the boundary conditions.

There are two important properties of the operator L that we need (Exercise 2):

- L is *self-adjoint* in the sense that $(Lu, v) = (u, Lv)$ for any $u, v \in \mathcal{H}$; and

- L is *positive* in the sense that $(Lu, u) > 0$ for any nonzero $u \in \mathcal{H}$.

It is worth noting that if L is a matrix, u and v are vectors, and $(\cdot, \cdot)$ is the usual Euclidean inner product, then self-adjoint and positive correspond to symmetric and positive definite, respectively (Exercise 3).

With these properties in hand, we can demonstrate the crucial duality mentioned earlier: solving the differential equation $Lu = f$ is *formally* equivalent to minimizing the functional

$$F(u) \equiv \frac{1}{2}(Lu, u) - (f, u)$$

over $u \in \mathcal{H}$. The functional F is *quadratic* in u and it has instructive scalar and matrix analogues (Exercises 4 and 5). The minimization of F is often written in the compact notation

$$u = \text{argmin}_{v \in \mathcal{H}} F(v), \tag{10.1}$$

which means "find the argument that minimizes F over all functions in $\mathcal{H}$." This link between the differential equation and the minimization problem lies at the heart of the finite element formulation. It is so important that we should study it for a moment to understand it.

Suppose u is a candidate function for minimizing F and $v \neq 0$ is any other function in $\mathcal{H}$. Consider the value of F at the point $u + v$. Using the linearity of L and the bilinearity of the inner product, we see that

$$
\begin{aligned}
F(u+v) &= \frac{1}{2}(L(u+v), u+v) - (f, u+v) \\
&= \frac{1}{2}\left((Lu, u) + (Lu, v) + (u, Lv) + (Lv, v)\right) - (f, u) - (f, v) \\
&= F(u) + \frac{1}{2}\left((Lu, v) + (u, Lv) + (Lv, v)\right) - \langle f, v \rangle.
\end{aligned}
$$

Because L is self-adjoint, we know that $(u, Lv) = (Lu, v)$. This allows us to simplify the above expression further:

$$
\begin{aligned}
F(u+v) &= F(u) + (Lu, v) - (f, v) + \frac{1}{2}(Lv, v) \\
&= F(u) + (Lu - f, v) + \underbrace{\frac{1}{2}(Lv, v)}_{\text{positive}}.
\end{aligned}
$$

The last term in this expression is positive because L is a positive operator. Therefore, we see that if $Lu = f$, then $F(u + v) \geq F(u)$ for all $v \in \mathcal{H}$, which means that u minimizes F over $\mathcal{H}$. Conversely, if u minimizes F, then it follows that $(Lu - f, v) = 0$ for all $v \in \mathcal{H}$, which means that

$$(Lu, v) = (f, v) \quad \text{for all } v \in \mathcal{H}. \tag{10.2}$$

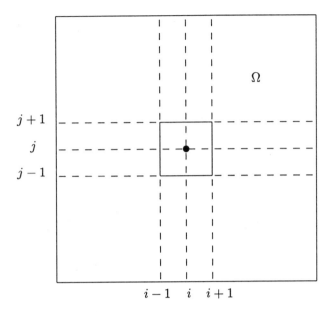

Figure 10.1: *A domain Ω showing the four elements surrounding grid point (i, j).*

This condition, which is useful for implementing finite elements, is essentially another way of saying that $Lu = f$. Thus, we have shown that the differential equation and minimization problem are essentially equivalent.

The finite element discretization of the problem can now be described. Suppose that the domain is the unit square, $\Omega = (0, 1) \times (0, 1)$, on which we place a uniform $(n + 1) \times (n + 1)$ grid, Ω^h, with mesh size $h = \frac{1}{n}$. As before, let (x_i, y_j) be the grid point with coordinates (ih, jh). However, we now focus not on the grid points, but on the square regions surrounding each grid point; these small regions are called *elements* (Fig. 10.1). Notice that there are four elements surrounding each grid point. The sets of four elements corresponding to neighboring grid points overlap in one or two elements. (Everything that follows can also be carried out on nonuniform grids.)

We begin with a common choice of approximating functions. Let H^h be the subspace of $\mathcal{H}$ consisting of *piecewise bilinear* functions u^h: each $u^h \in H^h$ is zero on $\partial\Omega$, continuous on Ω, and bilinear within each element. This means that a typical function in H^h has the form $u^h(x, y) = axy + bx + cy + d$ on each element, as depicted in Fig. 10.2. The level h discretization of (10.1) is

$$u^h = \mathrm{argmin}_{v^h \in H^h} F(v^h). \tag{10.3}$$

We have seen that the minimization problem on $\mathcal{H}$ is equivalent to solving (10.2) for $u \in \mathcal{H}$. This equivalence is also true on H^h. Thus, the discrete minimization problem (10.3) is equivalent to finding $u^h \in H^h$ so that

$$(Lu^h, v^h) = (f, v^h) \quad \text{for all } v^h \in H^h. \tag{10.4}$$

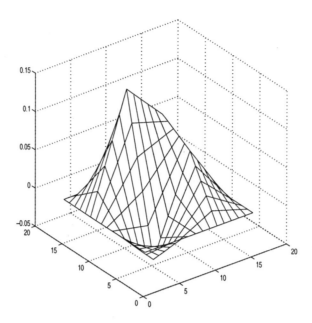

Figure 10.2: *A typical function in the space H^h used to approximate the solution to the problem. It is zero on $\partial\Omega$, continuous on Ω, and bilinear within each of the nine elements pictured.*

Finite Element Terminology. The terminology associated with finite element methods is not entirely standard and can get confusing. The discretization of the minimization problem is generally called the *Rayleigh–Ritz* formulation. The finite element method also has a *Galerkin* form, which starts not from the minimization problem (10.1), but from the equation $Lu = f$. As with Rayleigh–Ritz, the idea behind the Galerkin form is to approximate the solution $u \in \mathcal{H}$ by $u^h \in H^h$. However, the Galerkin approach requires the residual $Lu^h - f$ to be orthogonal to every $v^h \in H^h$. This condition also leads to (10.4). Thus, when L is a positive, self-adjoint, linear operator, and F is quadratic, the Rayleigh–Ritz and Galerkin formulations are essentially the same.

The practical solution of (10.4) requires several additional considerations. First, we need to choose a *local* basis for H^h consisting of functions, each of which is nonzero on its own particular patch of four elements. Specifically, let $\epsilon_{i,j}^h(x, y)$ denote the piecewise bilinear function in H^h that is centered on an interior grid point (x_i, y_j); it has the value 1 at (x_i, y_j) and is zero at all other grid points (Fig. 10.3). Any $u^h \in H^h$ can then be expanded as

$$u^h(x, y) = \sum_{i,j=1}^{n-1} u_{i,j}^h \epsilon_{i,j}^h(x, y). \tag{10.5}$$

The *nodal value* $u_{i,j}^h$ gives the value of u^h at (x_i, y_j); for this reason, (10.5) is called a *nodal basis expansion*.

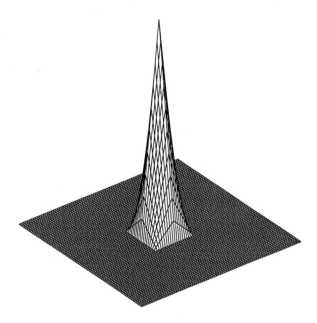

Figure 10.3: *A typical basis function $\epsilon_{i,j}^h(x,y)$ for the space H^h. It is nonzero, only on the patch of four elements that have (x_i, y_j) as a node.*

The next step is to substitute expansion (10.5) into expression (10.4). However, this task leads to a paradox: the operator L is second order and the approximating function u^h is linear in x and y, so $Lu^h = 0$ in each element! How can Lu^h be zero when we know generally that $(Lu^h, u^h) > 0$? The answer is that Lu^h is *not* zero on Ω. In fact, Lu^h does not even make sense on Ω because u^h is not suitably smooth: the first partials of u^h are piecewise smooth and therefore square integrable, but their discontinuities at the element boundaries prevent second derivatives from being square integrable. This means that we really cannot substitute (10.5) into (10.4). Fortunately, there is a way out of this dilemma: we just rewrite (10.4) in a form that requires fewer derivatives. This is done by applying the Gauss divergence theorem (the analogue of integrating by parts) to (Lu, v), where $u, v \in \mathcal{H}$, and using the fact that u and v vanish on $\partial\Omega$ (Exercise 1):

$$
\begin{aligned}
(Lu, v) &= \int_\Omega (-u_{xx} - u_{yy})v \, d\Omega \\
&= \int_\Omega (u_x v_x + u_y v_y) \, d\Omega \\
&= (\nabla u, \nabla v).
\end{aligned}
$$

This maneuver replaces the second-order derivatives in the problem by first-order derivatives. Making this replacement in (10.4), we are led to the *weak form*

$$(\nabla u^h, \nabla v^h) = (f, v^h) \quad \text{for all } v^h \in H^h. \tag{10.6}$$

Strictly speaking, this new problem is more general or *weaker* than the differential equation: a solution may exist, but it may not be twice differentiable, which the classical solution of $Lu = f$ requires. We stay with the weak form for the remainder of the chapter, so we are now free to be specific about what *suitably smooth* means: $\mathcal{H} \subset L^2(\Omega)$ is assumed to consist of functions that vanish on $\partial\Omega$ and that have first partials in $L^2(\Omega)$.

With the weak form of the problem, we can now substitute the expansion (10.5) into (10.6). Furthermore, we choose the so-called *trial* functions v^h to be the basis functions $\epsilon_{i,j}$. The result of this substitution is a linear system whose unknowns are the nodal values $u_{i,j}^h$. The matrix coefficients in this system are inner products of the form $(\nabla\epsilon_{i,j}^h, \nabla\epsilon_{k,\ell}^h)$ and the right-side values are $(f, \epsilon_{i,j}^h)$. Note that with our chosen bilinear basis functions, only neighboring basis functions overlap; thus, the inner products $(\nabla\epsilon_{i,j}^h, \nabla\epsilon_{k,\ell}^h)$ are zero unless $k = i$ or $i \pm 1$ and $\ell = j$ or $j \pm 1$. Evaluating the nine inner products associated with the (i, j) patch results in a row of the *stiffness matrix* (Exercise 6) given by the stencil

$$A_{i,j}^h = \frac{1}{3}\begin{pmatrix} -1 & -1 & -1 \\ -1 & 8 & -1 \\ -1 & -1 & -1 \end{pmatrix}. \tag{10.7}$$

This is the just the nine-point finite difference stencil scaled by $\frac{1}{3}$ (instead of the usual $\frac{1}{3h^2}$).

The inner products $(f, \epsilon_{i,j}^h)$ involve the integrals

$$(f, \epsilon_{i,j}^h) = \int_{x_{i-1}}^{x_{i+1}} \int_{y_{i-1}}^{y_{i+1}} f\epsilon_{i,j}^h \, dxdy,$$

which are generally approximated numerically. The simplest numerical integration scheme amounts to replacing the function f by its value $f(x_i, y_j)$ (Exercise 7):

$$\begin{aligned} (f, \epsilon_{i,j}^h) &= \int_{x_{i-1}}^{x_{i+1}} \int_{y_{i-1}}^{y_{i+1}} f\epsilon_{i,j}^h \, dxdy \\ &\approx f(x_i, y_j) \int_{x_{i-1}}^{x_{i+1}} \int_{y_{i-1}}^{y_{i+1}} \epsilon_{i,j}^h \, dxdy \\ &= \frac{h^2}{4} f(x_i, y_j). \end{aligned} \tag{10.8}$$

When all rows of stiffness matrix (10.7) and corresponding right sides (10.8) are assembled, the result is the matrix equation

$$A^h \mathbf{u}^h = \mathbf{f}^h. \tag{10.9}$$

Here we have introduced the source vector $(\mathbf{f}^h)_{i,j} = \left(\frac{h^2}{4} f(x_i, y_j)\right)$ and the solution vector $(\mathbf{u}^h)_{i,j} = \left(u_{i,j}^h\right) \in \Omega^h$.

This discretization shows that finite elements and finite differences boil down to similar matrix problems. But it is important to keep in mind that the methodology used to obtain them is quite different. Practical implementation of a multigrid scheme for finite elements can be done easily if we are guided by—*and remain faithful to*—the basic principle of functional minimization. For example, it may seem convenient to rescale (10.9) by dividing both sides by h^2, but this would change the relationship between the various grids used in the multigrid scheme, and it could easily introduce conceptual errors into the overall process.

We develop the multigrid scheme in the abstract by focusing on the functions u^h that solve (10.3) and (10.4), rather than their nodal values $\mathbf{u}^h$. First consider relaxation, whose goal is to provide an inexpensive method for eliminating oscillatory errors in the approximation, v^h. We can do this by making local changes of the form

$$v^h \leftarrow v^h - s\epsilon^h_{i,j}, \tag{10.10}$$

where $s \in \mathbf{R}$ is a suitably chosen step size. But how should s be chosen? Our plan of being faithful to the minimization principle gives us the answer: choose the best step size in the sense that it minimizes the functional over all possible choices. The mathematical statement is

$$s = \operatorname{argmin}_{t \in \mathbf{R}} F(v^h - t\epsilon^h_{i,j}). \tag{10.11}$$

This gives us the following relaxation scheme, which amounts to using (10.10) and (10.11) to sweep over the grid points:

For each $i, j = 1, 2, ..., n - 1$:

compute $s = \operatorname{argmin}_{t \in \mathbf{R}} F(v^h - t\epsilon^h_{i,j})$ and make the replacement $v^h \leftarrow v^h - s\epsilon^h_{i,j}$.

For our special functional F, this *coordinate relaxation* scheme is none other than Gauss–Seidel applied to (10.7) (Exercise 8).

The coarse-grid correction process is also easy to formulate in the abstract. We define the coarse-grid space $H^{2h} \subset H^h$ as the set of piecewise bilinear functions associated with the standard coarse grid Ω^{2h} formed by deleting the odd-numbered lines of Ω^h. The goal is to change the approximation v^h by a function $v^{2h} \in H^{2h}$ that approximates the presumably smooth error. The form of this correction is $v^h \leftarrow v^h + v^{2h}$. But how should v^{2h} be chosen? Again, being faithful to the minimization principle gives us the answer: choose the best coarse-grid correction in the sense that it minimizes the functional over all possible choices. The mathematical statement is

$$v^{2h} = \operatorname{argmin}_{w^{2h} \in H^{2h}} F(v^h + w^{2h}). \tag{10.12}$$

The coarse-grid correction scheme is

Compute $v^{2h} = \operatorname{argmin}_{w^{2h} \in H^{2h}} F(v^h + w^{2h})$ and set $v^h \leftarrow v^h + v^{2h}$.

This correction step, together with the coordinate relaxation scheme defined above, constitutes the core of the multigrid method. We can design practical algorithms based on this core just as we did in Chapter 3.

Our development of the finite element coarse-grid correction step appears simpler than it was for finite differences, because it comes naturally from the minimization principle. However, this simplicity is somewhat deceptive because it must still be expressed in terms of nodal vectors, and this requires intergrid transfer and coarse-grid operators. We will see that there are no choices that need to be made here: these operators are determined by the spaces and bases that we have already selected.

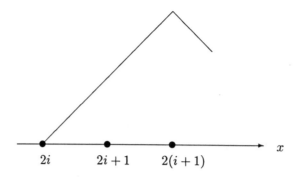

Figure 10.4: *Cross-section of the grid along a line of constant y showing the coarse-grid points $(2i, 2j)$ and $(2(i+1), 2j)$ and the intermediate fine-grid point $(2i+1, 2j)$. The coarse-grid element is linear in x between the two coarse-grid points.*

Consider first the process of adding an Ω^{2h} function, $v^{2h} \in H^{2h}$, to an Ω^h function. This process requires that we take the Ω^{2h} nodal representation for v^{2h},

$$v^{2h}(x, y) = \sum_{i,j=1}^{\frac{n}{2}-1} v_{i,j}^{2h} \epsilon_{i,j}^{2h}(x, y),$$

and convert it to a Ω^h nodal representation,

$$v^{2h}(x, y) = \sum_{i,j=1}^{n} v_{i,j}^{h} \epsilon_{i,j}^{h}(x, y).$$

In other words, we must find the coefficients $v_{i,j}^h$ that allow us to write v^{2h} as a function in H^h. We know that we can do this because H^{2h} is a subset of H^h, so v^{2h} must be in H^h. What we are looking for here is really just the interpolation operator I_{2h}^h that allows us to determine $\mathbf{v}^h = I_{2h}^h \mathbf{v}^{2h}$.

Consider the grid Ω^{2h} node with indices (i, j) that is located at the point $(i(2h), j(2h))$. This point is also the grid Ω^h node $((2i)h, (2j)h)$ with indices $(2i, 2j)$. Remembering that the coefficients in the nodal representations are just the function values at the nodes, we have

$$v_{2i,2j}^h = v^{2h}((2i)h, (2j)h) = v^{2h}(i(2h), j(2h)) = v_{i,j}^{2h}.$$

This says that $v_{2i,2j}^h = v_{i,j}^{2h}$.

Consider now the fine-grid node at point $((2i+1)h, (2j)h)$ (see Fig. 10.4). This node is halfway between nodes $((2i)h, (2j)h)$ and $(2(i+1)h, (2j)h)$ along a horizontal grid line. This line segment is part of a single coarse-grid element in which v^{2h} is bilinear and y is constant; thus, v^{2h} must be linear in x along this line. This leads us to conclude that

$$v_{2i+1,2j}^h = \frac{1}{2} \left(v_{i,j}^{2h} + v_{i+1,j}^{2h} \right).$$

For similar reasons, we have

$$v_{2i,2j+1}^h = \frac{1}{2} \left(v_{i,j}^{2h} + v_{i,j+1}^{2h} \right)$$

and

$$v^h_{2i+1,2j} = \frac{1}{4}\left(v^{2h}_{i,j} + v^{2h}_{i+1,j} + v^{2h}_{i,j+1} + v^{2h}_{i+1,j+1}\right).$$

As you can see, we have just reinvented the interpolation operator I^h_{2h} introduced in Chapter 3.

The next task is to determine the coarse-grid version of the operator A^h, which will eventually be called A^{2h}. It is most easily discovered by working with the nodal vectors and translating the minimization principle into matrix terms. As we have seen, discrete minimization principle (10.3) is equivalent to discrete equation (10.4), which in turn is equivalent to matrix equation (10.9). Similar reasoning shows that matrix equation (10.9) is equivalent to the matrix minimization principle

$$\mathbf{u}^h = \mathrm{argmin}_{\mathbf{v}^h \in \Omega^h} F^h(\mathbf{v}^h), \tag{10.13}$$

where

$$F^h(\mathbf{v}^h) \equiv \frac{1}{2}(A^h\mathbf{v}^h, \mathbf{v}^h) - (\mathbf{f}^h, \mathbf{v}^h)$$

(Exercise 5). Furthermore, similar reasoning also shows that coarse-grid correction problem (10.12) is equivalent to the matrix problem

$$\mathbf{v}^{2h} = \mathrm{argmin}_{\mathbf{w}^{2h} \in \Omega^{2h}} F^h(\mathbf{v}^h + I^h_{2h}\mathbf{w}^{2h}).$$

We use this form of the coarse-grid correction problem to carry out an informative calculation. Using the properties of inner products, we have (read this one with a pencil and paper!)

$$
\begin{aligned}
&F^h(\mathbf{v}^h + I^h_{2h}\mathbf{w}^{2h}) \\
=\ &\frac{1}{2}\left(A^h\left(\mathbf{v}^h + I^h_{2h}\mathbf{w}^{2h}\right), \mathbf{v}^h + I^h_{2h}\mathbf{w}^{2h}\right) - \left(\mathbf{f}^h, \mathbf{v}^h + I^h_{2h}\mathbf{w}^{2h}\right) \\
=\ &\frac{1}{2}(A^h\mathbf{v}^h, \mathbf{v}^h) + \frac{1}{2}(A^h I^h_{2h}\mathbf{w}^{2h}, \mathbf{v}^h) + \frac{1}{2}(\mathbf{v}^h, A^h I^h_{2h}\mathbf{w}^{2h}) \\
&+ \frac{1}{2}(A^h I^h_{2h}\mathbf{w}^{2h}, I^h_{2h}\mathbf{w}^{2h}) - (\mathbf{f}^h, \mathbf{v}^h) - (\mathbf{f}^h, I^h_{2h}\mathbf{w}^{2h}) \\
=\ &\frac{1}{2}(A^h\mathbf{v}^h, \mathbf{v}^h) - (\mathbf{f}^h, \mathbf{v}^h) \\
&+ \frac{1}{2}(A^h I^h_{2h}\mathbf{w}^{2h}, I^h_{2h}\mathbf{w}^{2h}) + (\mathbf{v}^h, A^h I^h_{2h}\mathbf{w}^{2h}) - (\mathbf{f}^h, I^h_{2h}\mathbf{w}^{2h}) \\
=\ &F^h(\mathbf{v}^h) + \frac{1}{2}\left(\underbrace{\left(I^h_{2h}\right)^T A^h I^h_{2h}}_{A^{2h}}\mathbf{w}^{2h}, \mathbf{w}^{2h}\right) - \left(\underbrace{\left(I^h_{2h}\right)^T}_{I^{2h}_h}\left(\mathbf{f}^h - A^h\mathbf{v}^h\right), \mathbf{w}^{2h}\right).
\end{aligned}
$$

Notice that the first underbraced item, $\left(I^h_{2h}\right)^T A^h I^h_{2h}$, is a matrix that operates on a coarse-grid vector. Furthermore, the role it plays in the inner product is analogous to the role played by A^h in the functional F^h. This suggests that we let $A^{2h} = \left(I^h_{2h}\right)^T A^h I^h_{2h}$, the coarse-grid version of A^h.

The second underbraced item, $\left(I^h_{2h}\right)^T$, must take a fine-grid vector into a coarse-grid vector; that is, it plays the role of a restriction operator. It therefore makes sense to let $I^{2h}_h = \left(I^h_{2h}\right)^T$.

With these operators identified, we may conclude the calculation. Letting $\mathbf{r}^h = \mathbf{f}^h - A^h\mathbf{v}^h$ and $\mathbf{f}^{2h} = \left(I_h^{2h}\right)\mathbf{r}^h$, we have

$$
\begin{aligned}
F^h(\mathbf{v}^h + I_{2h}^h\mathbf{w}^{2h}) &= F^h(\mathbf{v}^h) + \frac{1}{2}\left(A^{2h}\mathbf{w}^{2h}, \mathbf{w}^{2h}\right) - \left(\left(I_h^{2h}\right)\mathbf{r}^h, \mathbf{w}^{2h}\right) \\
&= F^h(\mathbf{v}^h) + F^{2h}(\mathbf{w}^{2h}).
\end{aligned}
$$

We should stand back and see what we have done. First, because $F^h(\mathbf{v}^h)$ is independent of $\mathbf{w}^{2h}$, the minimization of $F^h(\mathbf{v}^h + I_{2h}^h\mathbf{w}^{2h})$ is equivalent to the minimization of $F^{2h}(\mathbf{w}^{2h})$ over vectors $\mathbf{w}^{2h} \in \Omega^{2h}$; this is just the coarse-grid correction scheme introduced in Chapter 3.

Equally important, we have seen the variational properties of Chapter 5 emerge rather naturally through the minimization principle. They bear repeating:

$$
\begin{aligned}
A^{2h} &= I_h^{2h}A^hI_{2h}^h \quad \text{(Galerkin property)}, \\
I_h^{2h} &= \left(I_{2h}^h\right)^T.
\end{aligned}
$$

Thus, the finite element coarse-grid correction scheme amounts to choosing bilinear interpolation, its transpose for restriction, and the Galerkin property for determining the coarse-grid operator. Except for scaling differences, we have reinvented the transfer operators, coarse-grid operators, and the coarse-grid correction scheme developed in the previous chapters. It is important to keep in mind that for positive self-adjoint problems, we now have the option of developing multigrid solvers directly from the differential equations, as we did before, or using the minimization principle, as we did here. The differential equations may provide more flexibility simply because you do not have to adhere to optimality, but the choice to follow the minimization principle may give guidance and assurance that is less forthcoming from the differential equations.

Exercises

1. **Weak form.** For functions u, v that are sufficiently smooth on Ω and vanish on $\partial\Omega$, show that $(Lu, v) = (\nabla u, \nabla v)$. Hint: First apply the divergence theorem to the product $v\nabla \cdot \mathbf{u}$, where $\mathbf{u}$ is a vector-valued function and v is a scalar. Then remember that $Lu = -\Delta u = -\nabla \cdot (\nabla u)$.

2. **L is self-adjoint and positive.** Let u, v be sufficiently smooth functions that vanish on $\partial\Omega$. Show that the Poisson operator $L = -\frac{\partial^2}{\partial x^2} - \frac{\partial^2}{\partial y^2}$ is self-adjoint in the sense that $(Lu, v) = (u, Lv)$. Hint: Use Exercise 1. Show that L is positive in the sense that $(Lu, u) > 0$ for $u \neq 0$. Hint: Show that $(Lu, u) = 0$ implies u is constant.

3. **Self-adjoint and positive for matrices.** Show that if L is a matrix, u and v are vectors, and $(\cdot, \cdot)$ is the usual vector dot product, then self-adjoint and positive correspond to symmetric and positive definite, respectively.

4. **Scalar analogue.** Show that if $L > 0$ and f are scalars and $(\cdot, \cdot)$ is just scalar multiplication, then (Lu, u) is a quadratic function. Furthermore, $\frac{1}{2}(Lu, u) - (f, u)$ is minimized by the solution of $Lu = f$, which is $u = f/L$.

5. **Matrix analogue.** Suppose that L is a symmetric positive-definite $n \times n$ matrix, u and f are n-vectors, and $(\cdot, \cdot)$ is the usual vector dot product. Show that $\frac{1}{2}(Lu, u) - (f, u)$ is minimized by the solution of $Lu = f$, which is $u = L^{-1}f$.

6. **Stiffness matrices.** For fixed indices i and j, show that the inner products $(\nabla \epsilon_{i,j}^h, \nabla \epsilon_{k,\ell}^h)$ for $k = i$ or $i \pm 1$ and $\ell = j$ or $j \pm 1$ give the 3×3 stencil (10.7).

7. **Right-side inner products.** Show that

$$\int_{x_{i-1}}^{x_{i+1}} \int_{y_{i-1}}^{y_{i+1}} \epsilon_{i,j}^h f \, dx dy \approx \frac{h^2}{4} f(x_i, y_j),$$

where the region of integration is a single element.

8. **Gauss–Seidel.** Show that minimization problem (10.11) results in Gauss–Seidel applied to (10.9).

9. **Variational property.** Verify each of the steps in the simplification of $F^h(\mathbf{v}^h + I_{2h}^h \mathbf{w}^{2h})$ that led to the variational properties.

Bibliography

The ultimate resource for multigrid methods is the MG-Net website, located at
http://www.mgnet.org. The site is maintained by Craig Douglas and sponsored
by Yale University, CERFACS, the University of Kentucky, and the NSF. The
site features an extensive bibliography (over 3200 citations), links to free software (at least 22 packages at last count), information on conferences, and issues of
the *Multigrid Newsletter* (back to 1991). To subscribe to the newsletter, write to
mgnet@cs.yale.edu.

The Copper Mountain Conferences on Multigrid Methods have been held bi-
ennially since 1983. There have also been six European multigrid conferences.
Proceedings of recent conferences are available via MG-Net.

[1] R.E. BANK, *PLTMG: A Software Package for Solving Elliptic Partial Differential Equations. Users' Guide 7.0*, vol. 15, Frontiers in Applied Mathematics,
SIAM, Philadelphia, 1994.

[2] J.H. BRAMBLE, *Multigrid Methods*, vol. 294, Pitman Research Notes in Mathematical Sciences, Longman Scientific and Technical, Essex, England, 1993.

[3] A. BRANDT, *Multi-level adaptive solutions to boundary value problems*, Math.
Comput., 31, 1977, pp. 333–390.

[4] A. BRANDT, *Multigrid Techniques: 1984 Guide with Applications to Fluid
Dynamics*, GMD-Studien Nr. 85, Gesellschaft für Mathematik und Datenver-
arbeitung, St. Augustin, Bonn, 1984.

[5] A. BRANDT, *Algebraic multigrid theory: The symmetric case*, Appl. Math.
Comput., 19, 1986, pp. 23–56.

[6] A. BRANDT, S.F. MCCORMICK, J. RUGE, *Algebraic multigrid (AMG) for
sparse matrix equations*, in Sparsity and its Applications, D.J. Evans, ed.,
Cambridge University Press, Cambridge, UK, 1984, pp. 257–284.

[7] J.W. DEMMEL, *Applied Numerical Linear Algebra*, SIAM, Philadelphia, 1997.

[8] J.E. DENNIS AND R.B. SCHNABEL, *Numerical Methods for Unconstrained Optimization and Nonlinear Equations*, vol. 16, Classics in Applied Mathematics,
SIAM, Philadelphia, 1996.

[9] G. GOLUB AND C. VAN LOAN, *Matrix Computations*, Johns Hopkins University Press, Baltimore, MD, 1996.

[10] W. HACKBUSCH, *Multigrid Methods and Applications*, vol. 4, Computational Mathematics, Springer-Verlag, Berlin, 1985.

[11] W. HACKBUSCH AND U. TROTTENBERG, *Multigrid Methods*, vol. 960, Lecture Notes in Mathematics, Springer-Verlag, Berlin, 1982.

[12] D. JESPERSEN, *Multigrid methods for partial differential equations*, in Studies in Numerical Analysis, vol. 24, Studies of Mathematics, MAA, Washington D.C., 1984, pp. 270–318.

[13] S.F. MCCORMICK, ED., *Multigrid Methods*, vol. 3, Frontiers in Applied Mathematics, SIAM, Philadelphia, 1987.

[14] S.F. MCCORMICK, *Multilevel Adaptive Methods for Partial Differential Equations*, vol. 6, Frontiers in Applied Mathematics, SIAM, Philadelphia, 1989.

[15] S.F. MCCORMICK, *Multilevel Projection Methods for Partial Differential Equations* , vol. 62, CBMS-NSF Regional Conference Series in Applied Mathematics, SIAM, Philadelphia, 1992.

[16] J.M. ORTEGA AND W.C. RHEINBOLDT, *Iterative Solution of Nonlinear Equations in Several Variables*, Academic Press, San Diego, 1970. Reprinted as vol. 30, Classics in Applied Mathematics, SIAM, Philadelphia, 2000.

[17] U. RÜDE, *Mathematical and Computational Techniques for Multilevel Adaptive Methods*, vol. 13, Frontiers in Applied Mathematics, SIAM, Philadelphia, 1993.

[18] J. W. RUGE AND K. STÜBEN, *Algebraic multigrid*, in Multigrid Methods, vol. 3, Frontiers in Applied Mathematics, S. F. McCormick, ed., SIAM, Philadelphia, 1987, pp. 73–130.

[19] V. V. SHAIDUROV, *Multigrid Methods for Finite Elements*, vol. 318, Mathematics and Its Applications, Kluwer, Dordrecht, 1995.

[20] G.W. STEWART, *Introduction to Matrix Computations*, Academic Press, New York, 1973.

[21] G. STRANG AND G. FIX, *An Analysis of the Finite Element Method*, Prentice-Hall, Englewood Cliffs, NJ, 1973.

[22] K. STÜBEN, *Algebraic Multigrid (AMG): An Introduction with Applications*, GMD-Studien Nr. 53, Gesellschaft für Mathematik und Datenveranbeitung, St. Augustin, Bonn, 1999.

[23] L.N. TREFETHEN AND D. BAU, *Numerical Linear Algebra*, SIAM, Philadelphia, 1997.

[24] R.S. VARGA, *Matrix Iterative Analysis*, Prentice-Hall, Englewood Cliffs, NJ, 1962.

[25] P. WESSELING, *An Introduction to Multigrid Methods*, John Wiley and Sons, Chichester, 1992.

[26] D. YOUNG, *Iterative Solution of Large Linear Systems*, Academic Press, New York, 1971.

Index